AF781592

The Language of Agriculture in English

English For Careers

The Language of Agriculture in English

Andrew Jenkins-Murphy

Contributing Editor:
William M. Edwards, Ph.D.
Agricultural Economist
Iowa State University

Regents Publishing Company, Inc.

Cover: Wheat harvest in Ashland, Va.
Courtesy U.S.D.A., Soil Conservation Service
Photo by Tim McCabe

Cover design by Suzanne Bennett

Illustrations by Bobbi Angell

Photo Research by Monique Peyrat

Copyright © 1981 by
Regents Publishing Company, Inc.

All rights reserved. No part of this book may be reproduced in any form without permission in writing from the publisher.

10 9 8 7 6 5 4 3 2

Published by
Regents Publishing Company, Inc.
2 Park Avenue
New York, N.Y. 10016

Printed in the United States of America
ISBN 0-88345-350-9

TABLE OF CONTENTS

FOREWORD v

UNIT ONE
History of Agriculture 1

UNIT TWO
Energy and Equipment 13

UNIT THREE
Science of Agriculture 29

UNIT FOUR
Crops Around the World 43

UNIT FIVE
Farm Technology 59

UNIT SIX
Animal Agriculture 71

UNIT SEVEN
Business of Agriculture 85

UNIT EIGHT
Future of Agriculture 95

INDEX OF SPECIAL TERMS 107

FOREWORD

This book is one of a series of texts called *English for Careers*. The series is intended to introduce students of English as a foreign language to the language of different professional and vocational fields. The career areas that are covered are those in which English is widely used throughout the world, such as air travel, computer technology, the petroleum industry, atomic energy, and, in the case of this book, agriculture. None of these books is intended as a detailed training manual, but rather as a broad introduction to the subject.

Each book in the series serves several purposes. The first is to give the student an introduction to the particular vocational area in which he or she is interested. The type of work and the difficulties encountered in the field are discussed. The many areas of research and practice of agriculture around the world are described.

From the point of view of teaching English as a foreign language, these books are intended for a student at the high intermediate or advanced level. The student who uses these books should be acquainted with most of the structural patterns of English. The principal goals of the learner should be mastering vocabulary, using the various patterns in a normal mixture, and improving ability to communicate in English.

Each unit begins with a glossary of special terms in which words and expressions used in the field of agriculture are defined. An index of these terms is also presented in the back of this book. The glossary is followed by a vocabulary study which tests the student's comprehension of the special terms and gives practice in their use. In the reading that follows, these terms are used again within a contextual frame of reference. Each reading is followed by questions for comprehension and discussion. They give the student the opportunity to use in a communicative situation both the vocabulary items and the structural patterns encountered in the reading.

Each unit ends with a review section including exercises. Some of them pose situations that might occur if the student were actually working in the field. In doing these exercises, the student will also practice the specialized vocabulary of the field and other new words, as well as the structural patterns that are used with them.

A great deal of successful language learning comes from experiences in which the learning is largely unconscious. In offering these books, it is hoped that the student's interest in his or her chosen field will increase his or her ability to communicate more effectively in English.

I wish to thank the staff of the United States Department of Agriculture Experimental Station, Beltsville, Maryland for their assistance, and Jacqueline Flamm, Eugene J. Hall, and Carol F. Akiyama for their encouragement and editorial aid in this project.

Andrew Jenkins-Murphy
Washington, D.C.

The Language of Agriculture in English

UNIT ONE
A HISTORY OF AGRICULTURE

Special Terms

Agriculture: The practice or art of growing crops or raising animals useful to human beings.
Farming: The occupation of growing crops or raising animals. A *farmer* is a person who operates a farm or cultivates the land.
Arable: Land which is naturally suited for farming.
Domesticate: To adapt a plant or animal from its natural state to one more advantageous to human beings, usually under controlled conditions of growth and harvest.
Crop: A vegetable product of the soil or the total amount of a particular product grown in a season.
Cultivate: To prepare or work the land in order to raise food or some other product. *Tilling* the soil is cultivation by means of turning over the earth.
Harvest: The gathering of ripened crops.
Yield: The quantity or amount of a crop produced in a certain season.
Sow: To scatter or plant seeds in the ground.
Cutting: A root, stem, or leaf cut from a plant for the purpose of replanting.
Irrigation: The artificial watering of land.
Terracing: An alteration of the slope of a hillside for farming purposes. Horizontal or sloping ridges are built to conserve moisture or minimize erosion of the soil.
Fertile: Capable of producing abundant vegetation or crops.

Reap: To harvest, particularly grain. Until the 19th century, grain was reaped with a sickle, but in many parts of the world today it is done by machinery.
Silt: The fine, earthy sand usually carried by rivers and deposited as sediment.
Thresh: To beat the stems and husks of grain or cereal plants in order to separate kernels from straw.
Fallow: Arable land left idle for at least one season so that it can preserve moisture and accumulate nutrients.
Fodder: Crops grown as feed for farm animals.

Vocabulary Practice

1. What is *agriculture?*

2. What does the word *farming* mean? What is a *farmer?*

3. What is *arable* land used for?

4. How do farmers *domesticate* plants and animals?

5. What is a *crop*?

6. What does *harvest* mean?

7. What is the quantity or amount of a crop produced in a single season called?

8. How does a farmer *sow* seeds?

9. Explain the meaning of *reap* and name some ways it is done.

10. What does *cultivate* mean? What is cultivation by means of turning over the soil called?

11. What is *irrigation*?

12. What is a *cutting*? What is it used for?

13. Define *silt* and explain where it comes from.

14. What is the purpose of *terracing*?

15. What is the purpose of *threshing*?

16. Explain what *fallow* land means. What are its benefits?

17. What are *fodder* crops?

History of Agriculture

Agriculture is the world's oldest, largest, and most important industry. People depend on *farming* not only for food but for the raw materials of clothing, dyes, and medicines. Every country is vitally concerned with food, and its growth and distribution. The amount of food grown is of prime interest throughout the world.

About 11% of the earth's land is *arable*, or capable of producing crops without much modification. Another 22% is meadow or pastureland, suitable for raising animals. Tending these lands demands the energies of more people than all other occupations combined. More than one-half of the earth's population lives or works on farms.

For thousands of years our ancestors hunted animals and fish and gathered food that grew naturally—herbs, grains, nuts, and fruits. The earth's population was limited by its food supply, so people lived in small groups where food was available. Then, after the last great ice sheets melted about ten thousand years ago, people learned to *domesticate* animals and plants and the transition to a settled agrarian life began. Agriculture allowed people to shape their environment to suit their needs. *Crops* and animals were *cultivated* for use as food and clothing. This selection and cultivation greatly altered the balance of the earth's natural plant and animal life.

When our ancestors *harvested* wild foods, they learned to store a portion for periods of scarcity. Slowly, they began to adapt nature to their convenience. Some of the earliest tribes in western North America increased the *yield* of wild grass by building rock and earth barriers, or checkdams, to slow the runoff of water. The grass then seeded itself—one step short of human planting.

When people in various parts of the world learned that the land could be made more productive, permanent communities were established. *Sowing* seeds and planting *cuttings* permitted a division of labor which eventually led to society as we know it today. As our ancestors altered their environment, their population grew. The original small, primitive, local bands of people developed into nations. *Farming*—feeding people by cultivating crops—was essential to this development. The extensive written records left by the Greeks and Romans describe an elaborate system of farming which obviously was the product of centuries of trial and error.

While there is no one place to which the origins of agriculture can be traced, the earliest abundant evidence of the change from hunting to *tilling* is found in the northern hills and plains surrounding the Fertile Crescent, a region of southwest Asia extending in an arc from the eastern coast of the Mediterranean Sea across present-day Syria and Iraq to the northern coast of the Persian Gulf. Sometime between 10,000 B.C. and 9000 B.C., the cave dwellers in the region bordering the drainage basin of the Tigris and Euphrates Rivers began to settle in more permanent sites or hunting villages, intensified their collection of wild food, and began to experiment with agriculture. Sickles which date from 9000 B.C. have been found in parts of Palestine, but it is not known if the grain harvested was wild or sown. Wheat and barley, and possibly peas and lentils, were cultivated in large farming villages on the border of present-day Iran and Iraq. Recently, experiments with einkorn wheat—a short, flat, one-grain wheat commonly found in western Asia—have shown that a prehistoric family, working steadily for three weeks during a normal harvest, could have produced about a year's supply of clean grain.

By 6000 B.C., wheat and lentils were grown on Crete. Agriculture and its more settled way of life had reached eastward to Afghanistan, westward to Macedonia or Greece, and downward into the river plains of the Tigris and Euphrates in the south. From there, knowledge of farming spread along the Nile in Africa, northward into Europe through the Danube River, and along the Mediterranean coast.

There are indications that beans, peas, and water chestnuts were grown in northern Thailand by 7000 B.C., and rice may have been cultivated in the same region earlier. Millet, a small-seed cereal grass, was cultivated in China by 5000 B.C.

Archeologists study, interpret, and describe the material remains of our past cultures. In this way they have established that plant domestication in the Americas developed independently of that in Southeast Asia.

Courtesy the New York State Museum

A display depicting Seneca Indians harvesting corn.

In the Americas, pumpkins, squash, and beans in domesticated form were grown by at least 6000 B.C. The major cultivation centers were in middle America, the Amazon River basin, and the Andean highlands. In Mexico, *irrigation*, *terracing*, and the construction of islands in lakes increased land usage, but agriculture developed slowly there and occurred in widely dispersed areas of relatively poor *fertility*. Maize, or corn, was domesticated about 3500 B. C. In Peru, from about 2500 B.C. until the sixteenth century A.D., the chief crop was the potato. The Incas of Peru offer an example of a civilization adapting its agricultural processes to its environment. Because the days were hot and the nights cold all year long, the ancient Peruvians developed preserves of freeze-dried meat, fish, and mealy tubers, such as potatoes and manioc, that remains edible indefinitely. Giant warehouses lined the roadways in this rugged, mountainous part of the world.

Notable advances were made in Egypt after the basic ideas of agriculture spread from southwestern Asia sometime before 5000 B.C. Ancient Egyptian tombs and monuments show cultivation, seeding, harvesting with slotted *reaping* knives, the use of the Nile River for irrigation, and bread-making. Many plants and grains from the Middle East were readily adaptable to the highly productive Egyptian soil, replenished annually by *silt* from the flooded Nile. The records show extensive varieties of root crops as well as legumes, melons, olives, grapes, figs, and dates. A sophisticated society developed in Egypt and cities sprang up to serve as marketplaces.

By about 4500 B.C., the people of Mesopotamia were processing flax fibers into linen thread and making oil from flax seeds. In contrast to Egyptian cities, theirs were centers from which the inhabitants left to work in the fields. In what is now southern Iraq, the people initiated irrigation on the Tigris-Euphrates plain. Prior to this, farmers relied on natural flooding and rainfall for irrigation and could not plan for the future with any degree of certainty.

As irrigation systems expanded, rules for using the water and for keeping the systems coordinated and in good repair led to legal and political integration over large areas. Political stability encouraged permanent living arrangements. One such example is southwestern Arabia, where the Great Ma'rib Dam was built around 600 B.C. Another is the Negev Desert in southern Palestine, which was cultivated by means of terracing and irrigation, and which later became one of the breadbaskets of the Roman Empire. With the fall of Rome, the Negev returned to its semi-arid condition and is only now being reclaimed for agriculture.

Several major advances in agriculture took place between the years 400 and 1000 A.D. The countries of Europe had evolved a system in which people farmed for landlords who owned the land.

The peasants, or country people, worked the land around small communities in a one-field, one-crop system. Seed was sown at random and grain was reaped with a sickle. The grain was then *threshed* by large animals treading on it. This system developed into a plan which allowed a different piece of land to lie *fallow* each season. In the two-field system, one field would be planted with grain while the other would "rest." The three-field system was similar; two fields would be planted while one lay fallow. The three-field system also added the element of rotating crops in the strips of land. The rotations most often used in Europe during this era were: one strip of winter rye or wheat; one strip of spring barley, oats, peas, beans, or mixed corn; and one strip fallow. This system not only protected the soil, but allowed moisture to accumulate for a season. The fallow strip grew weeds which were burned at harvest time.

Horses and oxen were fitted with wheels which made them efficient workers. The horse collar, probably invented in China, became one of the most important developments of the time. It replaced the old harness band that pressed on the animal's windpipe, restricting its pulling power. The new padded collar enabled horses to pull heavier loads and to labor longer.

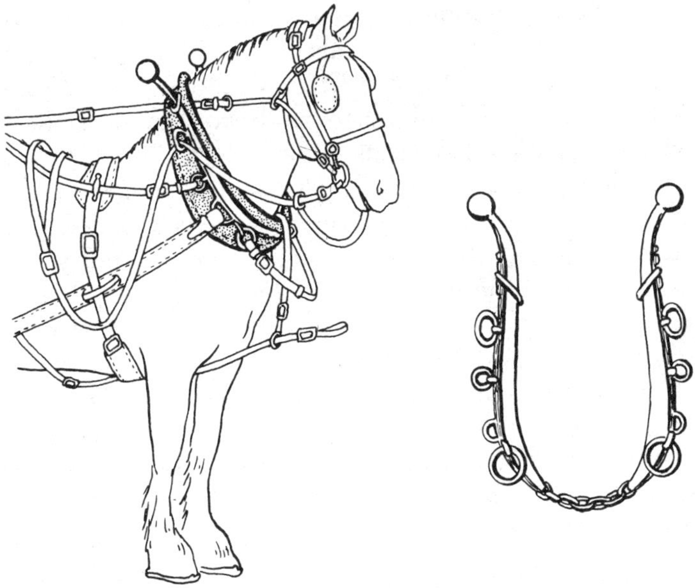

The padded horse collar allows the animal to pull heavy loads for long periods of time.

Courtesy FAO

Workers harvest coffee at a United Nations-sponsored agricultural project in the Wollamo subprovince of Ethiopia.

Colonization and exploration changed world agriculture. Seeds, plants and farm techniques were exchanged and, as contacts multiplied, the rate of advancement accelerated. Potatoes—later to become the mainstay of Irish and northern European peasants—tobacco, maize, and many other crops were brought from the new world to the old. Each wave of colonizers brought new devices and numerous crops, some of which transformed entire economies. Coffee is a good example: At the start of the sixteenth century, the coffee-growing centers of the world were Ethiopia, western Arabia, and Somaliland (now part of Somalia), where the crop grew wild on the fringes of the deserts. By the nineteenth century, the colonized lands of Java, Malaya, India, and Brazil had become major coffee producers. In much of Asia, Africa, Latin America, and Oceania there was a gradual enrichment of local farming with new crops and farming methods developed by European nations. The plantation system, in which each farm concentrated on one crop, also developed during this time.

By the time of the eighteenth century industrial and scientific revolution, agriculture was widespread. Only those few peoples on the fringes of civilization still survived by means of hunting and gathering. Over the past two hundred years, advances have come in several areas. New farm machinery has been manufactured and old implements improved, so that fewer people are needed to produce more food. New seeds and cultivation methods have been developed. New land use ideas have evolved, such as the Norfolk four-course system, developed in England, making it unnecessary to leave fields fallow. This system places emphasis on *fodder* crops which not only provide food for farm animals but help enrich the soil in which they are grown, thus increasing subsequent yields.

Discussion

1. What portion of the world's population is engaged in agriculture?

2. What percentage of the world's land is used for growing crops?

3. How did people feed themselves before the development of agriculture?

4. When did people first begin to raise crops?

5. Why did our early ancestors store some of the food they had hunted or gathered?

6. Why was agriculture important to population growth?

7. Where is the Fertile Crescent? What do we know about agriculture there?

8. Name some of the earliest crops grown in the Middle East.

9. How far had agricultural knowledge spread from the Fertile Crescent by 6000 B.C.?

10. What do archeologists do?

11. What is known about agriculture in ancient Thailand?

12. Where were the major cultivation centers of the Americas?

13. How was land usage increased in Mexico?

14. How did the Incas deal with their severe climate problems?

15. What were some of the products grown in ancient Egypt?

16. Where was flax introduced? When? What were its uses?

17. How did the cities of Mesopotamia contrast with those of Egypt?

18. Why is the Negev Desert interesting and important agriculturally?

19. What does fallow mean? Why was it an important technique in farming?

20. What is the two-field system? The three-field system? Give an example of a rotation of crops in the three-field system.

21. Give an agricultural example of the effect of colonization on a country's economy.

22. Where were the centers of coffee-growing in the sixteenth century?

23. Are crops raised only as food for humans?

24. What is the Norfolk four-course system? Where was it developed?

Exercises

A. Complete the following sentences with the appropriate word or phrase:

1. Agriculture is the ______________ , ______________ , and ______________ industry in the world.

2. In prehistoric times, people gathered wild foods such as _______ , ______________ , and ______________ .

3. People first learned to domesticate plants and animals _______ years ago.

4. Western North American Indians built __________ to slow the runoff of ______________ .

5. The people of ancient Palestine used ______________ to harvest grain around __________ B.C.

6. Two important rivers in the area called the Fertile Crescent are the ______________ and the ______________ .

7. By 6000 B.C., ______________ and ______________ were cultivated on the island of Crete.

8. ______________ study the material remains of our past cultures.

9. A small-seed cereal grass called ______________ was grown in China around 5000 B.C.

10. Name five crops grown in the Americas in ancient times: 1) _______, 2) _______, 3) _______ , 4) _______ , 5) _______ .

11. Egypt's soil was made fertile by the annual deposit of __________ caused by the flooding of the __________ River.

12. Flax is used for __________ and __________ .

13. Rules for controlling water usage became necessary as _________ systems developed.

14. The Great Ma'rib Dam was built in _______ B.C. in __________ .

15. The Negev Desert was once so fertile that it was called the _________ of the Roman Empire.

16. Between 400 and 1000 A.D., a new system developed which called for a strip of land to lie __________ for a season.

17. The three-field system contributed the idea of ____________ the crops to be planted.

18. Potatoes were the mainstay of the ___________ and ______________ peasants.

19. By the nineteenth century, Brazil had become a major __________ producer.

20. By 1900, agriculture had replaced __________ and ___________ over most of the world.

21. The Norfolk four-course system placed a new emphasis on __________ crops.

B. Explain how agriculture affected ancient civilizations. What was the relationship between the growth of cities and agriculture?

C. What are some agricultural problems in your country?

D. Describe some major turning points in the history of agriculture. What was the importance of each one?

UNIT TWO
ENERGY AND EQUIPMENT

Special Terms

Draft animals: Strong animals used to pull heavy loads on farms.
Roughage: Coarse food which is difficult for human beings to digest; fodder or forage is roughage fed to domesticated farm animals.
Grain: Seeds grown as a food source.
Milling: Grinding or crushing grain into a fine flour. The device used for this purpose is called a *mill;* the earliest types of mills were hand grinders called *querns.*
Weeding: Removing undesirable plants called *weeds.*
Tractor: A vehicle with a chain-belt tread or large rear wheels, used for pulling plows, harrows, reapers, and other farm machines.
Hoe: An implement with a thin, flat blade on a long handle, used for weeding or loosening the earth around plants.
Plow (or **plough**): An instrument used to cut, lift, and turn over soil, especially for preparing a seedbed. A *plowshare* is the front part which cuts the furrow in the soil; a *moldboard* is the part just behind the plowshare which lifts, turns, and pulverizes the soil.
Harrow: A cultivating implement with spike-like teeth or discs, used for pulverizing and smoothing the soil.
Roller: A revolving cylinder used to pack soil and crush clods.
Acre: A unit of land measurement in the United States and Great Britain, equivalent to 1/640 of a square mile or 4840 square yards. A *hectare* in the metric system of measurement is equal to 10,000 square meters or about 2½ acres.
Furrow: A trench in the soil, usually made by a plow.

Fertilizer: A natural or synthetic material which supplies nutrients to help plants grow.
Mow: To cut grasses with a *sickle* or a machine.
Reaper: A tool or machine which cuts and gathers the grain harvest.
Flail: An instrument for threshing grain by hand, consisting of a wooden handle at the end of which a thicker, shorter stick swings freely to beat the grain.
Winnow: To separate grain from chaff by tossing it into the wind.
Combine: A harvesting machine which performs the operations of reaping, threshing, and cleaning grain while moving over a field.

Vocabulary Practice

1. What are *draft animals* used for on farms?
2. Why is *roughage* fed mainly to animals rather than humans?
3. What is *grain*?
4. Explain the use of a *mill.* What was the earliest type called?
5. How are plants which are harmful to crops removed from the soil?
6. What kinds of jobs do *tractors* do on a farm?
7. What is a *hoe* used for?
8. What is a *plow*? What are some of its parts?
9. What is a *harrow*?
10. What is a *roller* used for?
11. How many *acres* are in a *hectare*? Where is the unit of measure called an acre used? What units of measure are used in your country?

12. What are trenches dug by plows called?

13. Why do farmers put *fertilizer* on their land?

14. What is the meaning of *mow*?

15. What is a *reaper* used for?

16. Describe a *flail*. What is it used for?

17. Why do farmers *winnow* grain?

18. What can a *combine* do?

Energy and Equipment

For several thousand years after people began to cultivate the soil, human energy was used for all farm work. The muscle power of the farmers themselves supplied the energy to till, plant, and harvest the fields. The Indians in the Americas were still farming land in this manner when Columbus arrived in the late fifteenth century, fully 8000 years after the earliest known cultivation of plants in the Western Hemisphere.

Adapting the energies of animals for agricultural tasks was a major advancement. Cattle were domesticated by farmers in southwestern Asia by 5000 B.C., and horses were used for transportation around the Persian Gulf by 3500 B.C. But it was probably not until around 3000 B.C. that *draft animals* capable of pulling heavy loads were used to help till the soil.

These beasts of burden—usually water buffaloes, horses, oxen, and camels—got a large part of their nourishment from foods normally inedible by humans, such as grass, alfalfa, clover, and hay. This meant that farmers were able to convert previously unusable *roughage* into an energy form which eased human labor.

By the fifth century B.C., animals were used for the agricultural task of *milling*. Until this time, dried *grains* were ground by a hand instrument called a *quern*. With the invention of the rotary quern, animals were brought into the flour-producing phase of farming. Cen-

Courtesy Netherlands Consulate General

A typical windmill in Heusden, the Netherlands.

turies later, the Romans introduced rotary *mills* powered by the force of flowing water. As this idea spread, people in many parts of the world began to use water power as energy for other kinds of agricultural activities.

In the tenth century A.D., the concept of wind power was introduced in Persia, and lowland countries like Holland quickly adopted the idea of the windmill. Both water and wind power were of limited use, however, because dry and windless seasons brought work to a standstill.

About 5000 years after the harnessing of draft animals, a series of discoveries allowed farmers to make use of additional sources of energy. In the nineteenth century, the invention of the steam-powered engine, and later the internal combustion engine, were major breakthroughs. Steam power is the force of vapor from boiling water, used to operate machines. Many devices were adapted to this type of energy, but most were large and clumsy. They soon gave way to internal combustion engines which used the heat made by burning fuel to produce motion. This added mechanical power based on fossil fuels to human and animal energies. It quickly became apparent that these machines could be adapted to all facets of agricultural production. In much of the world today, particularly where rice and wheat are important crops, most of the sowing, cultivating, *weeding*, and harvesting is still done by hand or with the help of animals. Many countries, however, rely almost totally on machine-driven tools and devices developed in the twentieth century. These machines are considered essential to all phases of farming, from preparing the soil for seeds to processing, loading, and delivering the final produce.

Courtesy Caterpillar Tractor Company

The earliest tractors were powered by steam and traveled at about three miles per hour.

Courtesy Caterpillar Tractor Company

The Clark Farm in Pullman, Washington, uses a modern Caterpillar brand tractor to pull a disc plow through tough grass.

Perhaps the most important mechanical development in the field of agriculture during this century is the *tractor.* First came the steam tractor, burning coal or straw. Although it was clumsy and bulky, a small one had the pulling power of about twenty horses. Factory production of gasoline-burning tractors started in 1903. By the 1930s, power lifts and hydraulic or liquid-forced lifts were added, and today these internal combustion traction engines have virtually replaced other energy sources in much of the industrialized world. Tractors, some with several hundred horsepower, are efficient for cultivating, sowing, reaping, pumping water, and hauling.

EQUIPMENT FOR PREPARING THE SOIL OR SEEDBED

Farmers must have discovered early that plants grow better when the ground is broken and the soil turned over. At first they used sturdy, pointed digging sticks, often weighted with stones, to turn the soil in small plots. Next came the hand *hoe,* originally a blade of stone or hard wood fastened at an angle to a long handle. *Plows,* used in

Courtesy USDA Soil Conservation Service

A large wheel tractor pulls a modern plow to prepare a barley field for fall replanting.

Mesopotamia and Egypt by 3000 B.C., were designed to be pulled in long, straight rows. When metals were added, broad hoes could be used in stony soil; some were strong enough to cut through the roots of shrubs and trees. Iron *plowshares* up to eighteen inches long were in use in Britain in 300 B.C. When draft animal power was introduced, larger and more complex plows were designed in such a way that animals could pull the full weight of the implement. The evolution of the plow, however, was a gradual one until the eighteenth century when Jethro Tull of England designed lighter but stronger tools. Some modern plows have mold boards which roll the soil over and bury weeds growing on the surface. Others, called disc or chisel plows, loosen the soil but leave more plant material on the surface to reduce soil erosion.

Harrows are plows used primarily for stony or rocky fields. These A-shaped implements with hard wood or iron teeth were used extensively by the Romans to break up clods, smooth the ground, cover seeds, and uproot weeds. Disc harrows, with as many as twenty sharp, flat, circular metal plates, have been used in Japan for hundreds of years to loosen the soil and kill weeds.

Another major seedbed-preparation implement is the *roller*. Originally, farmers dragged heavy tree limbs over the land; later, animals and then tractors were used to pull logs or weighted cylinders. Rollers with either ridged or smooth surfaces are used to crush clods and to pack the soil to help preserve moisture.

The basic ideas of ground-breaking have remained the same throughout the ages. Some of the earliest types of plows are still in use in certain parts of the world, but modern plows are sturdy and efficient and easily pulled by tractors. Today a mechanically well-equipped small farmer can plow three to eight *acres* a day. Some very large tractors, and implements can plow as much as 100 acres in a single day.

Courtesy USDA Soil Conservation Service

A seed drill is used to sow grass between rows of seven-year-old apple trees at Symms' Fruit Ranch near Greenleaf, Idaho.

EQUIPMENT FOR PLANTING SEEDS

In the early days of agriculture, seeds were either sown broadcast—tossed in all directions—or dropped by hand into shallow holes, one at a time. Around 1500 B.C., the Assyrians developed a seed planter which dropped grain or small seeds through a tube mounted on a plow into *furrows* made by the plow. Somehow this drill-planting method was lost and did not reappear until the invention of Jethro Tull's horsedrawn seed drill and cultivator in the eighteenth century. Both implements made a hole, dropped a metered amount of seed at a measured distance, and turned the earth over it. In the past two hundred years, many variations of the seed drill have appeared. Some use a drag chain or wheel to pull a certain amount of soil over the seed, while others place a small amount of *fertilizer* beside the seed. Multiple-row drills use a system of force feeding through fluted or gouged wheels. These wheels, driven by gears from their axles, or main shafts, govern the amount of seed dropped into the tube.

Types of planting machines are as numerous as types of crops. Airplanes have even been used to broadcast rice. Transplanting machines set out plants that have sprouted in another bed; they open a furrow, pour in a small amount of water, set the plants with rubber "fingers," and press down the soil around them.

Generally, it is best to disturb the soil as little as possible while planting. Most tropical crops are surface-rooting; tea, cocoa, rubber, bananas, coffee, citrus, oil palms, and coconuts suffer when their surface roots are damaged by excessive cultivation. But in some areas, such as the maize-growing region of Thailand, the seed planters must deposit seeds at least ten inches deep. In these cases the soil must be loosened to allow new seedlings to push through to the surface.

Most seeds benefit from fertilization. Modern machinery can shred manure and spread it evenly. There are also broadcast spreaders for granular synthetic fertilizers. Corn and cotton present a special problem. Weeds spring up easily between the rows, sapping the plants' food and absorbing moisture. Seeding machines for these crops can loosen the soil between the rows and spray it with a weed-controlling chemical. Some crops, usually those planted in rows, are cultivated after the plants have emerged so that weeds can be removed.

In some cases, the same machine is capable of cultivating and loosening the soil—with small shovels, rotary hoes, or discs—and depositing the seeds. In southeast Asia there are foot- or machine-powered devices which water and seed rice paddies simultaneously.

EQUIPMENT FOR HARVESTING

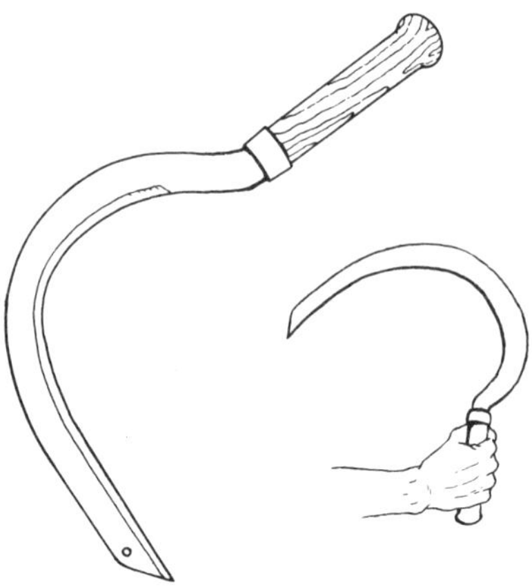

The *sickle*—an implement which, more than any other, symbolizes the agricultural laborer around the world—has been in use for at least ten thousand years. Farmers developed this tool to *mow* wheat; it grabs the stalks at the top and cuts them with quick, short strokes. The earliest sickles were used in the Middle East and were made of flint blades set in wooden or bone handles. Scythes, which are similar to sickles, have been used at least since the time of the Roman Empire. These are somewhat more efficient than sickles but are still hand tools like the grain cradle, a scythe with rows of wooden fingers above the cutting edge to catch the severed portion of the grass.

Between the first and fourth centuries, the Gauls, in what is present-day France, developed a *reaper* which could be used on level ground. The device was pushed by a donkey or an ox through a field of grain. The heads of the stalks were ripped off by a single long blade and fell into a box. The tool worked well but, like the Assyrian seed planter, it was lost and therefore not used by other cultures. Mechanical reaping did not reappear until the eighteenth century.

After cutting and gathering, the next step in making grain usable is threshing. Farmers in ancient Egypt and Asia threshed their grain by spreading it on a hard floor and driving oxen over it. The Romans pulled a sledge studded with sharp flints over the threshing floor. Others beat it with a jointed, wooden-handled, L-shaped stick called a *flail.* After threshing, loose grains were separated, or *winnowed,* from the chaff and straw by sifting them through large sieves or by tossing them into a breeze so that the heavier grain kernels fell directly to the ground while the chaff blew away.

During the eighteenth and nineteenth centuries, mechanical reapers and threshers were introduced, greatly simplifying these processes. In the 1830s in the United States, Obed Hussey and Cyrus McCormick designed reapers which cut and raked, using several blades mounted on a cylinder which spun while moving across the field. One driver with two draft horses pulling a reaper could harvest fifteen acres of wheat a day instead of the half-acre a farmer could cut by hand. Further experiments led to the development of the *combine,* the most advanced labor-saving device for grain farming. Today this machine, which reaps, threshes, and winnows grain in one operation, is self-propelled or drawn by a tractor; some models can harvest an acre in eight minutes.

The concept of combining harvesting operations in the field using one machine spread quickly. Rice combines now use oversized tires with long mud lugs, or protrusions, in soft or wet fields. There are peanut combines as well as large complicated machines that harvest sugar cane and lift the stalks into trucks. Modern harvesters for crops such as potatoes and sugar beets pull the tubers from the ground, shake off the dirt, and load them onto a wagon; some even bag the potatoes. There are also feed-grinding machines to cut and finely chop such crops as grass, green corn, and sorghum. Modern corn-pickers perform several tasks: the ears are picked, the machine husks or removes the outer covering, the shelling attachments and blowers remove the kernels from the cob, dropping the shelled corn into a bin.

Other machines have been developed for harvesting crops such as hay. The hay is mowed first then sun-dried, collected, and used to feed animals. There are a number of available machines: a hay-crusher cuts or crimps stems to speed drying, a raker gathers the hay into rows, a baler presses it tightly and loads the bales onto wagons. All operations are accomplished by the use of powerful gasoline or diesel engines mounted on the machines themselves or contained in the tractor pulling the machine.

SPECIAL EQUIPMENT

Some mechanical devices have stimulated whole new industries and greatly expanded others. The development of mechanical refrigeration in 1890 meant that agricultural products could be transported over longer distances. The invention of the cotton gin and the flax-processing machine opened large markets for these products. Today there are cotton-picking machines which gin or strip the usable cotton from the seeds right in the field.

The farm equipment in use today is adapted to special environmental and social conditions. Japan, with a high labor-to-land ratio, is applying mechanical power to raise the yield rather than save labor.

Courtesy USDA Soil Conservation Service

A self-propelled combine reaps, threshes, and winnows oats in one operation.

Japanese farmers have the highest per acre yields in the world in many product areas. They have adapted the internal combustion engine to small farms by developing two-wheeled power tillers called garden tractors. In some parts of the United States, hydroelectric power is used for pumping water, grinding and mixing feed, and heating farms. Nuclear energy is used in some areas to supply electricity and may one day actually operate field equipment.

Common in parts of the world with high humidity or short harvesting seasons is the artificial drying of crops, particularly grain, in large, hot-air blowing chambers run by electricity, gas or diesel fuel. Because it allows earlier harvesting, this method reduces the risk of crop loss to insects or bad weather. The high cost of these energy sources, however, is encouraging people to look for other, more economical sources of energy.

Most of the technology for bringing mechanical power to farms comes from the industrialized countries of the world; only recently have these resources become available to agricultural societies in quantity and manner sufficient to affect production. Despite these advances, about half the world's farmland is currently plowed by draft animals. Most farmers in Nepal, other scattered areas of Asia, Latin America, and Africa still rely on human labor alone. In Brazil all three major forms of energy are widely employed: Large commercial farmers use tractors and other mechanical implements, small farmers rely on draft animal power, and Indians in the Amazon basin work the land entirely by hand.

Despite increased technology and production, large-scale farm equipment is just not suitable for use in many parts of the world. Ownership of only small machines is economical where farms are small in size. Even then, machines may be jointly owned by more than one family. Human or animal power may be the only practical approach to farming where fields are steep and uneven, or soils very wet. In areas where labor is much less costly than machinery and fuel, mechanization on any scale is uneconomical.

The use of farm machinery on hard soils or in the application of important chemicals may increase agricultural production. In some cases, the use of machinery is merely a labor-saving device with no yield advantage. And in the harvesting of certain crops, human labor may actually result in less crop damage than with mechanization.

Discussion

1. What was the first source of energy used by the early farmers?

2. When were animals first used for draft power? Give some examples of animals used in this way.

3. What types of food do draft animals eat? Why is this important?

4. After the introduction of draft animals, what were the next major advances in energy or power sources for the farmer? Why did these prove to be unsatisfactory?

5. What were the disadvantages of these new sources of power?

6. What was the significance of the tractor to farmers? What were the first types of tractors? What characteristics do modern tractors have?

7. When were internal combustion engine tractors first mass-produced?

8. What are digging sticks? How were they used?

9. Why were metal hoes an improvement over those of stone or wood?

10. What are harrows used for? Describe those used by the Romans. What refinement did the Japanese add to the harrow?

11. What is the purpose of plowing the soil?

12. What were the earliest rollers made of? How is a roller used?

13. What is the reason for packing the soil?

14. Give examples of some tools used for tilling.

15. What was the earliest method of planting seeds?

16. What is the similarity between the ancient Assyrian grain planter and the one invented by Jethro Tull more than three thousand years later?

17. Describe the operation of a transplanting machine.

18. What special jobs do seeding machines accomplish in corn and cotton fields?

19. How long have farmers been using sickles? What does one look like?

20. What are scythes? How are they used?

21. How did the grain cradle work?

22. How did Egyptian farmers thresh their grain?

23. How did the machines invented by Hussey and McCormick improve grain harvesting?

24. What is a combine? Name several types of combines.

25. How do corn-pickers work?

26. Why is grain dried artificially?

27. What are feed-grinding machines?

28. What was the significance of the invention of the cotton gin?

29. What are garden tractors? Where are they used?

30. When was mechanical refrigeration developed?

31. Give some examples of the uses of hydroelectric power on farms.

32. Name a country where most of the farming is done without animals or machinery.

33. Describe some conditions under which large-scale farm equipment cannot be used.

Review

A. Look at the pictures on the left. On the right, identify each object and give a brief description of its uses.

1. 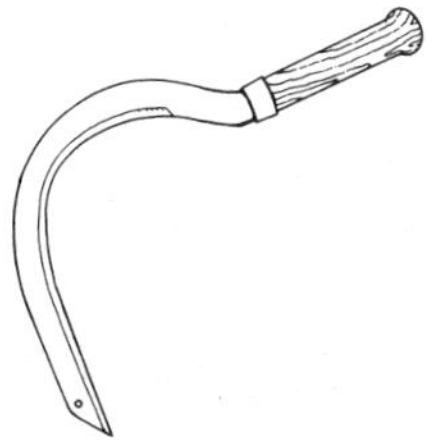

____

____

2.  ______________________
____

B. Match the short definitions in Column A with the words in Column B:

Column A	Column B	
a) animal feed	harrow	_____
b) a draft animal	seed drill	_____
c) spike-toothed, A-shaped machine	flail	_____
d) a soil-packing, heavy cylinder	McCormick	_____
e) horse- or machine-drawn planting machine	combine	_____
	gin	_____
f) strong traction engine machine	fodder	_____
g) L-shaped threshing stick	roller	_____
h) inventor of a reaper	fertilizer	_____
i) compact bundle of hay	ox	_____
j) machine for stripping seeds from cotton	tractor	_____
k) machine which reaps, threshes, and winnows grain		
l) plant food		

C. What kind of energy source is most common in your country? Are machines widely used? If so, for which crops? What special machines are most needed?

D. Describe one complete agricultural process, such as growing wheat, from preparation of the soil to the final product.

E. Compare the uses of draft animals and machines in agricultural production.

UNIT THREE
SCIENCE IN AGRICULTURE

Special Terms

Hybrids: The offspring of two different species or of different varieties of the same species.

Genetics: The study of the heredity and variation of organisms. As an applied science, genetics aims toward improving a particular strain, breed, or variety of plant or animals.

Variety trials: Genetic experiments, usually conducted on special farms, which test different strains of a plant and attempt to improve its quality or adaptability.

Soil: Material formed biochemically from weathered or decomposed rock and organic matter, covering much of the earth's surface and capable of supporting vegetation.

Nutrition: The process that sustains or supports life and growth in living things. A *nutrient* is an ingredient which nourishes a living thing.

Erosion: The natural process by which the earth's surface is worn away by wind and water. It is most severe on sloping or exposed land.

Leaching: The washing away of harmful or nutritive elements by water which seeps through the soil. It is often severe in areas of high rainfall.

Elements: The fundamental or basic parts of all matter, divided into 103 chemically named units.

Macronutrients: Those elements in soil which are essential in relatively large amounts for the well-being of plants: nitrogen, phosphorus, potassium.

Micronutrients: Those elements needed in small or trace amounts to support plant life: manganese, boron, chlorine, sodium, zinc, cobalt, copper, molybdenum. Other necessary elements labeled neither macro- nor micro-nutrients are: sulfur, magnesium, calcium, iron.

Organic: Related to or containing living organisms. *Inorganic* refers to mineral matter other than plant or animal. The organic portion of soil, which has been formed by the decomposition of vegetable or animal matter, is called *humus.*

Legume: A plant capable of taking nitrogen from the air and using it as fertilizer for the soil.

Manure: The excretion of animals used as fertilizer. It is applied to soil to increase its organic matter content and to provide nutrients.

Pesticide: A general term for substances used to kill insects, rodents, weeds, and other pests.

Herbicide: A substance used to destroy weeds.

Fungicide: A substance used to destroy fungi.

Insecticide: A substance used to kill insects.

Vocabulary Practice

1. What is a *hybrid* plant or animal?
2. Why is *genetics* an important part of agriculture science?
3. Why are *variety trials* conducted?
4. What is the function of *soil*?
5. What are nourishing ingredients in living organisms called?
6. What is *nutrition*?
7. What is *erosion*? Where does it most often occur?
8. What is *leaching* and where does it usually occur?

9. What are chemical *elements*? Name some *macronutrients*. Name some *micronutrients*.

10. What is the difference between *organic* and *inorganic* substances?

11. What is *humus*?

12. Where do *legume* plants obtain nitrogen?

13. How is *manure* used? Where does it come from?

14. What is a *pesticide*? An *herbicide*? A *fungicide*? An *insecticide*?

Science in Agriculture

Methods of farming and food production advanced slowly and steadily over several thousand years. In the middle of the nineteenth century, progress in agricultural developments accelerated as scientists began to study plants. Experiments in crossing or mixing two existing plants produced *hybrids*. The conditions necessary for plant growth were better understood and it became possible to control those growth processes by altering nature.

The science of *genetics* began in the 1860s in Austria with the work of Gregor Johann Mendel, who experimented with peas, and with Luther Burbank of the United States, who studied potatoes. Scientists learned how the physical characteristics of plants were passed on to their offspring. From this knowledge, Charles Darwin of England began to develop hybrids which would adapt to the soil of a particular region.

By the early part of the twentieth century, agricultural research and experimentation was widespread and the results of these scientific efforts were being applied in the fields. Farmers in different parts of the world were able to apply knowledge from the laboratory to their special problems.

Most agricultural research is conducted by formally organized experimental stations, colleges, and universities. This research is

Courtesy FAO

A team of researchers from the UN Food and Agriculture Organization inspects a cooperative coffee farm in Mogiana, Brazil.

usually financed and administered by international agencies or by national or local governments. Large sums of money, in the form of grants to research institutions, are spent by private industry, business groups, and farmers' associations. Private companies also conduct agricultural research themselves, usually to develop new marketing products. In many countries, the publicly administered experimental stations conduct programs on private farms in cooperation with farmers. These programs serve many purposes. Frequently they provide a forum for demonstrating the values of new, improved practices. They may be simple *variety trials* or they may be attempts to determine if techniques developed elsewhere can be adapted to local conditions.

Japan, with only 5% as much cultivated land as the United States, has almost the same number of agricultural experimental stations. These two countries support different types of stations. In the United States, most research is carried on at multi-purpose stations with facilities designed to deal with a broad range of regional agricultural activities. In Japan, most of the stations have a single or limited purpose and deal with a single crop, a small group of crops, or a single characteristic of a crop.

Over the past several decades, genetic research has developed many new varieties of crops. Various types of research are now being conducted. The most successful are in the areas of *soil,* fertilizers, *nutrition,* and weed- and pest-controlling agents. A by-product of the research concerning the basic *nutrients* for plants and animals is its contribution to the treatment of nutritional deficiencies and diseases in humans.

SOIL, FERTILIZERS, AND NUTRIENTS

A thin layer of soil covers much of the earth's surface. Varying in thickness from a few inches to a few feet, soil is where the plant, animal, and mineral worlds meet in a dynamic relationship. Plants grow by using valuable nutrients and water from soil; animals depend on plants for food; the decomposed residue from both regenerates the soil so that the cycle can start again.

To be conducive to good plant growth, soil must be porous enough for roots to penetrate it easily. It must have temperatures in the proper range for biological activity, a water-storage capacity large enough to support plants between rains or irrigation periods, and enough connecting air pores to permit exchange of oxygen for carbon dioxide through the roots' metabolism.

It takes nature several hundred years to build an inch of topsoil, but *erosion, leaching,* poor crop rotation, or poor fertilization can render soil unproductive within months. Farmers and scientists are constantly working to conserve soil in its most life-supporting and fertile state. Several methods are in use around the world to help stop erosion and leaching. Trees or hedgerows are planted along shelter lines to check the wind. Deep-rooted thickly growing crops are planted to hold the soil in place and to keep the rain from washing it away. Terraces are built on sloping land to hold the soil in place or catch it as the rain washes it down the hillside. Contour plowing, in which the furrows run across the slope of the land rather than with it, keeps water from running off too quickly. Planting alternating strips of row crops and soil-holding crops also reduces erosion.

Soil must provide all the essential *elements* needed for plant growth. While plants receive carbon, hydrogen, and oxygen from the sun, air, and water, the earth must provide vital elements in mineral form. Soil contains nearly all the mineral elements in nature, some in great quantities, others in very small amounts. The elements gener-

Courtesy USDA, photo by Erwin Cole

A Monticello, Missouri, farmer spreads fertilizer on his field. The field will next be plowed and planted.

ally required in the greatest amount by plants are nitrogen, phosphorus, and potassium; these are known as *macronutrients*. The next most frequently needed elements are sulfur, magnesium, calcium, and iron. Other elements such as manganese, boron, chlorine, sodium, zinc, cobalt, copper, and molybdenum are needed in tiny amounts; these are called trace elements or *micronutrients*.

Fertilizers are necessary wherever land is cultivated. Plants that grow naturally in a region do so because they have chosen that area as the best one for their needs. When soil is plowed, seeded and cultivated, plants foreign to the area grow there by force, causing aeration and faster decomposition of the natural organic matter; the natural flora is removed and the cycle of revolving nutrients is broken. Thus, even the richest natural soil will grow poor if fertilizers are never added. Once the nutritional needs of a plant are understood, farmers can add fertilizers to replace or supplement those absorbed into the plant.

An important step forward in understanding plant nutrition was made in 1840 by Justus von Liebig, who advanced a theory based on his analytical investigations of the ashes of plants and their mineral contents. He demonstrated that the *inorganic* elements of soil, extracted by plants, could be replaced by artificially produced inorganic fertilizers. Liebig and others found that soil also contains *humus*, or *organic* matter vital for producing life. Some of this matter originated with algae or dead bacteria, the residue of a time when the earth was completely covered with water.

Tropical soils are almost always deficient in macronutrients due to heavy rainfall and large quantities must be added in order to produce reasonable crops. Most soils contain enough micronutrients for field crops but special fertilizers must be added for certain vegetables and fruits. It is important to determine how much and what type of fertilizer to add.

Nitrogen is an especially important fertilizer, not only because it is so readily absorbed (one ton of harvested wheat removes fifty pounds of nitrogen from the soil), but also because it greatly increases yields of sugar cane, maize (corn), millet, rice, sorghum, and yams when added in the proper form. Experiments in the central United States, for example, have repeatedly demonstrated that the addition of one hundred and sixty pounds of nitrogen per acre of corn produces almost twice the yield of that achieved when no additional nitrogen is applied. Certain plants, called *legumes*, can take in nitrogen directly from the air through nodules found on their roots, and thus do not require additional nitrogen as fertilizer.

TYPES OF FERTILIZERS

ORGANIC

Manure: The most abundant natural fertilizer; an animal by-product which contains large amounts of nitrogen and potassium.

Green manure: Live plants plowed under to decay. Legumes such as clover and alfalfa trap nitrogen from the air; when plowed under, they enrich nitrogen-depleted soil

Bone meal: Ground bone scraps which release many elements over a four- to five-year period; high in phosphorus.

Guano: The excrement of bats and sea birds, used mainly in the tropics.

Others: Dried fish scraps, dried blood from slaughterhouses, rotted compost or plant material, by-products from crop processing and sewage can add nutrients to soil.

INORGANIC

Lime: Used to overcome acidity and to add calcium and phosphorus.

Union Carbide photo issued by FAO

Ammonia: Common source of nitrogen. Applied as a liquid, or as a gas called anhydrous ammonia.
Phosphate rock: the chief source of phosphorus, formed originally from the bones of sea animals.
Potash and salt deposits: Mined for their potassium ammonia.
Synthetics: Manufactured fertilizers, from thousands of sources.

Plants "eat" food by absorbing minerals. The application of fertilizers to "feed" plants or soil takes varied forms. While a few plants can absorb fertilizers sprayed in weak solutions onto their leaves, most must do so through their roots. Commercial synethetic fertilizers come in the form of granules, pellets, powders, liquids, and gases. These, and the natural or organic fertilizers, may be applied by equipment attached to a tractor. They can be spread evenly across a field, deposited alongside seeds by a grain drill, or pumped into the soil by tubes attached to pressure tanks.

PROTECTING PLANTS

The protection of plants before, during, and after harvesting is an increasing challenge to agricultural scientists. Today's *herbicides* are highly effective weapons against wild plants which use up water and nutrients needed by the main crop. The chemicals destroy weeds but are neutralized by contact with the soil before they can damage the seeds of the main crop. *Fungicides,* which kill fungi that cause plant diseases both during growth and after harvesting, are also generally effective when applied properly. The biggest problem concerning plant protection is insect or pest control.

Insecticides often affect the delicate balance between the various plant and animal communities in even the smallest areas. While the control or elimination of a plant pest may immediately benefit the farmer, there may be longer-lasting effects which are harmful or dangerous. Some substances effectively destroy insects but leave toxic or poisonous residues in the crop, making it dangerous for consumption. Other insecticides alter an area's ecological cycle by destroying animals which prey on the target insect; this produces stronger pests which, having somehow survived the treatment, breed more rapidly because their natural enemies have been destroyed.

The value of insecticides is accepted, provided they are highly selective and interfere as little as possible with the network of life. The immense difficulties science faces in developing effective plant protection are complicated by the ability of many insects to develop resistance to chemicals in a relatively short time, and by the fact that there are many useful insects which need protection. Many fruit and vegetable crops are pollinated—their seeds activated for growth—by varieties of insects which might be affected by strong insecticides.

Organic insecticides are the most widely used but are considered persistent; they are long-lasting and affect other living things. Inorganic insecticides produced from minerals are less persistent. Botanical insecticides, such as nicotine from tobacco and dried pyrethium flowers, are effective against specific pests such as aphids. Bacterial insecticides fight pests by infecting them with diseases. Perhaps the most interesting approach to plant protection is biological; chemicals are used to sterilize insects, rendering them incapable of reproduction.

FAO photo by F. Mattioli

A Laotian farmer sprays seedlings at Dong Dok with pesticide from a sprayer provided by the UN Food and Agriculture Organization.

Courtesy USDA

Fungicide is applied to an orange grove near Lake Wales, Florida, by helicopter.

The serious necessity for pest control is emphasized in a study made by the United Nations Food and Agriculture Organization (FAO) estimating that, each year, insects alone destroy 10% of the world's production of stored cereals or grains. The development of *pesticides* has greatly increased our ability to produce food, and has reduced the difficult labor needed to combat weeds and insects. However, a great deal of additional research and testing is needed to fully understand their long-term effects on our total environment.

Discussion

1. Who was Mendel? Why was his work important? Name two other experimenters in the field of genetics.

2. Where is agricultural research often conducted?

3. How does private industry aid in agricultural research?

4. Contrast the experimental stations in the United States with those in Japan.

5. Describe the value of soil to the animal, vegetable, and mineral worlds.

6. What qualities must be present in soil for good plant growth?

7. List several methods of soil conservation used to reduce erosion and leaching. In what ways are they effective?

8. List the elements needed by plants for growth. Which come from the sun, air, and water? Which come from the soil?

9. Why are fertilizers used on cultivated land?

10. Why is the understanding of plant nutrition important?

11. Who was Justus von Liebig? What was his contribution to agricultural science?

12. Give some examples of crops which give larger yields when fertilized with nitrogen.

13. What are the main types of organic fertilizers? Which is the most abundant? Which releases its elements over several years?

14. What are the main types of inorganic fertilizers? Which is the chief source of phosphorus?

15. How do plants "eat"?

16. How are fertilizers applied to the soil?

17. Why are weeds injurious to crops?

18. What are some major problems surrounding the uses of insecticides?

19. List some ways to protect plants from pests.

20. What type of insecticide is most widely used? What are its drawbacks?

21. Do you think pesticides have been helpful to the world, or do they constitute a serious danger?

Review

A. Indicate whether each of the following statements is true or false:

1. _____ A hybrid is obtained by crossing or mixing two existing plants.
2. _____ Both Mendel and Burbank experimented with potatoes.
3. _____ Research conducted in experimental stations is useless to the average farmer.
4. _____ Experimental farms are sometimes used as forums for demonstrating new practices.
5. _____ Soil must be porous to allow roots to grow freely.
6. _____ Leaching is a type of fertilizer.
7. _____ Hedgerows are planted to help conserve soil.
8. _____ The elements most required by plants from the soil are nitrogen, phosphorus, and potassium.
9. _____ There are 103 macronutrients.
10. _____ Fertilizers help replenish soil which has been depleted of some nutrients.
11. _____ All fertilizers are inorganic.

12. _____ Some fertilizers are synthetic.
13. _____ Manure is an abundant fertilizer on some farms.
14. _____ Guano is the excrement of cattle.
15. _____ Fertilizers can be applied only by hand.
16. _____ Herbicides are universally dangerous to plants and soil.
17. _____ Some insects are valuable as seed pollinators.
18. _____ Botanical insecticides are manufactured from minerals.

B. What type of agricultural research is conducted in your country? Do you know how it is financed and administered?

C. What kind of fertilizer is most used in your region? Are there problems in finding the best fertilizer for your country's crops? How do you think the yields of these crops can be increased?

D. Discuss some major aspects of the uses of insecticides and other plant-protection techniques. What do you know of the effects of pesticides on your country's ecology—that is, the environmental relationship of the animal, vegetable, and mineral worlds?

UNIT FOUR
CROPS AROUND THE WORLD

Special Terms

Carbohydrates: A group of chemical compounds containing only carbon, hydrogen, and oxygen; sugar and starch are the most common examples.

Proteins: A group of complex organic nitrogen compounds containing amino acids; essential for growth and repair of human tissue.

Fats: A group of complex, high energy compounds found mostly in animal tissue.

Minerals: Natural, inorganic substances like calcium and phosphorus with chemical compositions that are valuable to human growth.

Vitamins: Various complex organic substances occurring naturally in plant, animal, and human tissue, and essential in small amounts for the control of the life process.

Annual: A plant which lives and grows for only one year or season.

Perennial: A plant with an indefinite life span; it recycles itself.

Drought: A long period of time with no rain, especially during a planting season.

Monsoon: The weather system that produces dry and wet seasons in India and southern Asia.

Livestock: Domestic animals such as cattle, horses, sheep, hogs, and goats, raised to provide food, clothing or transporation.

Grazing: Animal feeding on growing grasses in fields or on pastureland.

Rhizome: A root-like stem usually horizontal, growing under or along the ground that sends out roots from its lower surface and leaves or shoots from its upper surface.

Tuber: A swollen stem, usually underground, bearing buds from which new plant shoots arise; a potato is a common tuber.

Bulb: A modified underground stem, usually surrounded by scale-like leaves and containing food for the new plant stored within it; an onion is a common bulb.

Biennial: A plant that has a normal life cycle of two years or that requires two years to reach maturity, producing leaves in the first year, blooming and producing fruit in the second year, and then dying.

Photosynthesis: The process by which plants convert light, water and nutrients into energy for cell growth.

Greenhouse: A structure in which to cultivate plants that require controlled temperature and humidity; usually made of glass.

Vocabulary Practice

1. What elements do *carbohydrates* contain? Give examples of carbohydrates.
2. What do *proteins* contain? Why are they important?
3. What are *fats*?
4. Define *mineral*. Define *vitamin*. What is the difference between the two?
5. How does an *annual* plant differ from a *perennial*?
6. What is a *drought*?
7. Where does the phenomenon known as a *monsoon* occur? How does it affect that region of the world?
8. Give some examples of *livestock* and their uses.

9. What is *grazing*?

10. Describe a *rhizome*, a *tuber*, and a *bulb*.

11. Describe the life cycle of a *biennial* plant.

12. What is *photosynthesis*?

13. What is a *greenhouse* used for?

Crops Around the World

Considering that animals eat plants, it can be said that all of our food comes from plants. In a stricter sense, depending on culture and cost, 15% to 40% of the world's diet comes from animal products or animals, while 60% to 85% comes from plants. While there are perhaps a million species of plant life throughout the world, only about a hundred different main crops provide us with food, materials for clothing, and other useful products. By 1500 B.C., all the major food staples grown today were being cultivated. With the exception of tomatoes and coffee, we have domesticated no important new food plants in the last two thousand years.

GRAINS

Cereal, or grain, is the seed of plants such as wheat, rice, corn, rye, and oats. It is the largest and most important single food item throughout the world, requiring almost half the world's cropland. Grains may be cooked in their natural state or processed into breads, pastas, or beverages. All the great civilizations have been founded on grain crops. All the grains we value as food contain *carbohydrates*, *proteins*, *minerals*, *vitamins*, and a little *fat*. They offer a more balanced diet than any other single type of plant.

Grain plants are all species of grasses that once grew wild, but they differ in one important respect from the typical grasses seen in an open meadow: Grain crops are *annuals* which die when their life cycles are completed, and are replanted each season; pasture grasses are *perennials* which scatter to produce new plants on a continuing basis.

Several years ago, at the site of Jarmo, a 9000-year-old village in eastern Iraq, archeologists unearthed a number of charred grains of wheat almost identical to two types of wheat grown today. One was a wild species and the other a domesticated variety called "einkorn." Wheat is wider in geographic distribution than any other cultivated food crop and is the staple food for one-third of the world's population. Some 400 million acres of soil per year are used to grow wheat in almost every country from the equator north to the Arctic Circle. Wheat grows best in regions receiving less than thirty inches of rain a year; *drought* or heavy rainfall can dramatically affect its production. The world's leading wheat-growing nations are the Soviet Union, the United States, China, and Canada.

FAO photo by F. Mattioli

The women of Ma-Lu People's Commune, near Shanghai, China, weed a rice paddy.

Rice is the staple food of more than half the world's population. Thousands of varieties of rice have been developed, mainly in southern and southeastern Asia where the crop has been cultivated for at least five thousand years. Rice, unlike the other major cereals, grows best when it is in water for part of its life cycle. Thus, the principal rice-growing countries of the world lie in the *monsoon* belt where heavy seasonal rains permit farmers to flood large areas to a depth of several feet. World production of rice is about 400 million tons a year, one-third of which is produced by China and one-fifth by India.

Other principal producers are Pakistan, Vietnam, Japan, Indonesia, Thailand, Burma, Brazil, Egypt, and the United States. Rice is also a dietary staple in Central and South America. An acre of rice yields almost twice as much food energy as an acre of wheat.

Corn, which is indigenous to the Americas, is cultivated in much the same way as wheat but requires much more space and care. Corn can be grown only in regions where farmers can count on three or more frost-free months after planting. Corn contains less protein than wheat but is a staple in the diet of Central and South America and an important grain in other regions of the world. The great bulk of the world corn crop is used as feed for *livestock*, to supply industry with sugars and starches, and in whiskey production. The United States produces almost half the world's annual corn crop. Other major producers are Brazil, Mexico, Argentina, South Africa, India, and Indonesia.

Corn (*Zea mays*)

Rye, while less popular than wheat, rice, and corn, is important because it grows in poor, light soil that is unsuitable for other cereals. It is primarily a bread crop and is grown in cold countries such as the Soviet Union, Poland, and West Germany.

Europe produces the bulk of the world's barley and oat crop, while the grasses known collectively as millets and sorghum are grown principally in Africa and Asia. Each of these cereals is used for both human food and livestock fodder, but a large portion of the world's annual production of barley is used for brewing beer and distilling whiskey.

Teff is the staple grain of Ethiopia but is now widely cultivated elsewhere.

PASTURE PLANTS

Although human digestive systems cannot tolerate the fibrous stems and leaves of pasture grasses and clovers, these plants make up the principal diet of cattle, sheep, and many other animals which in turn provide us with meat and dairy products. Most pastureland used for *grazing* includes a range of grasses, legumes and herbs. Unlike the cereal crops which are cultivated for their seeds, pasture grasses are cultivated for their stems and leaves. Grass stems lose most of their food value for animals when they begin to develop seeds. The important pasture and meadow grasses are perennial ryegrass, Italian ryegrass, timothy, meadow fescue, cocksfoot, crested dogstail, buffalo grass, and Kentucky bluegrass, also known as smooth meadow grass.

Legumes are important to pasturelands because of their ability to "fix" atmospheric nitrogen in the soil with the help of bacteria living in their roots. They are the chief source of protein in pasturelands and complement the predominately carbohydrate diet of the grasses.

A mixture of grasses and clover in pastures can enrich the soil in ways which aid the growth of other crops. Farmers frequently grow clover just before planting grain. The legume second in importance to clover is alfalfa, which is extremely deep-rooting and strongly resistant to drought.

Sugar cane is the tallest of all economically important grasses, growing to a height of ten to twenty feet. It is an easy-to-grow perennial, requiring rich, deep, well-drained soil, a hot moist climate, an annual water supply of about sixty inches and regular application of nitrogen, potassium, and phosphate fertilizers. The growth period may be as short as eight months or as long as thirty months. Normal crop yields are about fifty-five tons of cane per acre, from which about five tons of sugar may be extracted. The major portion of the world's sugar cane crop is cut by hand, with a machete or another heavy, sharp knife. More than half the world's sugar supply comes from sugar cane. The principal cane-growing countries are Cuba, Brazil, and India.

FIBER PLANTS

Fibers occur in all but the simplest plants. They are made up of a special substance which gives pliant strength to the plant tissues, enabling even fragile stalks to bend without breaking in strong winds. The plant fibers most cultivated for use today are seed hairs from cotton and kapok, stem fibers from flax, jute, and hemp, and leaf fibers from sisal. Synthetic fibers have replaced many plant fibers in the clothing market, but natural fibers are still important for manufacturing paper, rugs, textiles, and other products. They constitute about 20% of the total value of the world's crops.

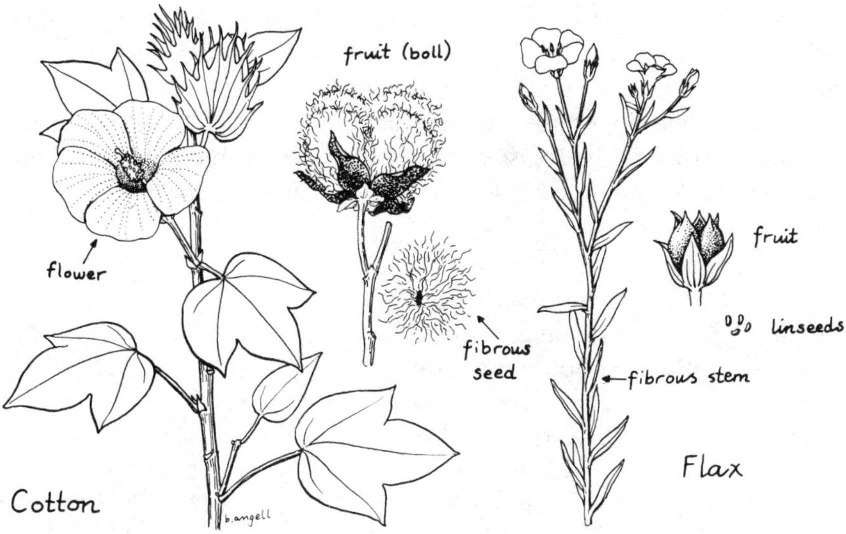

Cotton is the world's leading fiber for spinning thread and weaving cloth. It is used everywhere but is grown mainly in the United States, Egypt, Sudan, the Soviet Union, Brazil, India, China, Nigeria, South Africa, and the West Indies. The plant thrives best in a subtropical climate with spring rains and a dry, hot summer. Cotton is an annual, grows to a height of three feet, and is sown directly on well-tilled land in rows three feet apart. It needs careful weeding and irrigation, a great deal of fertilizer, and application of insecticides.

Kapok is the seed hair of a tree, often called the silk-cotton tree, that grows wild in tropical Africa and Asia and is cultivated on farms in Indonesia and the Philippines. It is used principally as a filler for mattresses and garments.

Flax, an annual plant, was used as long ago as 4000 B.C. for wrapping Egyptian mummies. It is raised in regions which have cool, moist summers followed by dry autumns. Flax produces a very strong thread or fabric. It is grown principally in western Europe, Canada, the United States, the Soviet Union, India, and Brazil.

Jute comes from a tropical plant that grows on the flat flood plains of the Ganges River around Calcutta, India, and in Bangladesh. It grows to a height of ten feet, yielding eight-foot fibers, and is used for making cheap, strong, wrapping cloths like burlap.

Hemp, which is native to Asia, makes strong, durable rope and is also the source of the drugs marijuana and hashish. Hemp has been replaced in many parts of the world by a stronger but lighter fiber called abaca or Manila hemp; this is native to the Philippines but now cultivated in Indonesia and Central America.

Sisal, a leaf fiber native to Mexico, is adapted to the dry climate on the edge of deserts. It also grows in East Africa, Indonesia, and the West Indies. Its main use is for packaging twine and for binder twine used in grain harvesters.

The seeds of some plants have a high oil content for which cooking oils and protein supplements for livestock feeds are derived. One example is soybeans, rapidly becoming an important world-market crop. The soybean is a legume that grows best in temperate climates with moderate rainfall. The United States, Brazil, Argentina, and China are important producers of soybeans.

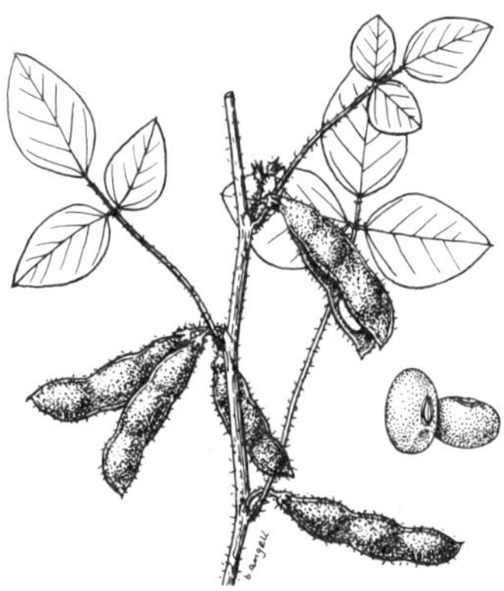

Soybeans *(Glycine max)*

Other important oilseed crops are flax (used in linseed oil), safflower, corn, cotton, sesame, peanuts, and African palm.

ROOTS, TUBERS, AND BULBS

Some plants propagate by a process known as vegetative reproduction, that is, by growing new stems, roots, or shoots that develop into new plants or extensions of the old ones. This process usually takes place under the surface of the soil with the development of *rhizomes* or *tubers*. The plants that develop *bulbs* are a less concentrated food form than grains and do not store easily, but they produce exceptionally high yields. Since they demand deep, heavy cultivation and usually increase the fertility of the soil, they are useful for crop rotation.

The turnip and rutabaga, which probably originated in western Asia and which grow wild along the banks of many European rivers, have been introduced to most parts of the world. These root plants are *biennial,* inexpensive and nutritious, and are used both as human food and animal feed. They are important in crop rotations that include wheat, pasture plants, and barley.

Sugar beets, sometimes called sea beets, thrive on fertile soils in lowlands with warm, sunny summers. They are green, flowered weeds with a root system rich in carbohydrates and sugar. They require large amounts of potash and water since each plant soaks up about thirty gallons of water during its growth period. They are biennial and produce about seventeen tons of beets per acre, from which about two tons of sugar can be extracted.

Onions, garlic, leeks, shallots, and chives—all biennials—are bulb plants. Their storage organs are not roots, but layers of leaf enclosing a small bud. The roots extend from the plant stem that forms at the base of the bulb; the layers of thin leaves protect the plant and help retain moisture within it. These plants are native to western Asia and the Mediterranean region, but are cultivated in many other parts of the world.

Tropical root crops are invariably perennials. They start life in the rainy season, concentrate their food reserves underground to survive the dry season, and come to fruition when the rains come again. The cassava or manioc plant is native to South and Central America but is now cultivated in every tropical region, notably West Africa and Malaysia. Yams, grown mainly in Africa, and sweet potatoes, native to Central America, are root plants which climb when staked; the roots are almost a foot long. They are only a little less nourishing than the potato. Taro, a tuberous root, is an important food on South Pacific islands. It is baked, boiled, or made into a porridge called "poi," and its leaves are eaten as a green vegetable.

Potatoes are indigenous to the moist and cool Andean highlands of Peru and Bolivia, but are now grown largely in the moist and cool lowlands of North America and northern Europe. The potato grows in almost any fertile soil except heavy clay or wet, undrained land. It is one of the easiest crops to raise, and high yields of good quality can be expected as long as the soil has been thoroughly cultivated and heavily fertilized with phosphate and potash. An average yield is nine tons

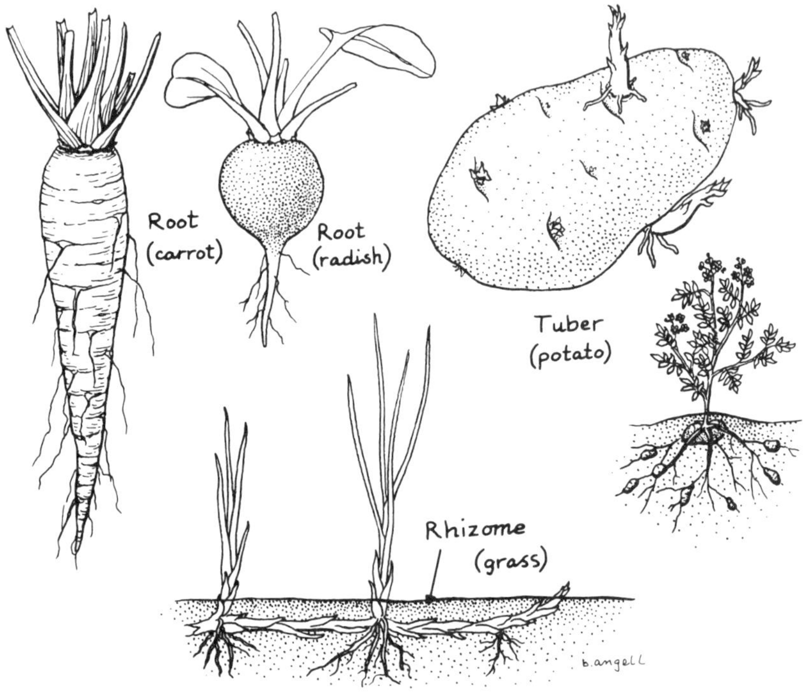

per acre. As a world food crop, the potato is only a little less important than wheat and rice. It is very nutritious, containing abundant carbohydrates, some protein, and small amounts of essential minerals. Like many other root crops, potatoes are indigestible unless cooked. Europe and the Soviet Union raise about three-fourths of the world's potatoes.

VEGETABLES AND FRUITS

Farmers throughout the world raise many kinds of vegetables, fruits, and nuts. Lettuce, native to the Mediterranean region, is cultivated everywhere. It is an annual which grows rapidly, storing its food reserves and water in its compact mass of leaves. Watercress, another green, grows year round in temperate climates, even when there is a slight frost. Cabbages are water-storing biennials which have been developed in varied forms around the world to fit the local climate: Kohlrabi is used mainly as fodder; Brussel sprouts are miniatures that can withstand frcst during growth; kale flourishes in mild winters; and cauliflower and broccoli, indigenous to the Middle East, thrive best in cool, moist climates. Asparagus is often etiolated, or kept from the sun to stop *photosynthesis*, to make its stalk more tender.

Soft-stemmed fruit and seed crops need fertile, thoroughly weeded soil and a great deal of sunshine. Many of the most common types are legumes, including the popular green pea. The pea is grown as both a field crop and a vine in the northern temperate zones; most of the world's pea crop comes from northern China. Other legumes include green beans, wax beans, butter beans, and many varieties of string beans. These are field crops and many can be dried and stored indefinitely. Some, like the kidney bean, are high in protein and are used as an inexpensive alternative to meat in parts of the Americas. Soybeans are today's most important source of plant protein in China and Japan.

The peanut, or ground nut, is the hard, ripe seed of a leguminous plant. It grows best in a subtropical climate and has value not only as a food, but as a base for paints and other products. Somewhat like peanuts, plants such as cucumbers, squash, pumpkins, and gourds have long trailing stems which enable them to absorb a great deal of sun.

Tomatoes and other annual soft fruits sprout, flower, and bear seeds only during hot, sunny summers. In cooler regions, *greenhouses* are used for spring planting. Tomatoes need rich soil and ample water to produce a nutritious product high in sugar and vitamins. Biennial soft fruits such as raspberries, blackberries, and strawberries seem to lead a double life. Each summer they develop tall stems, then bear flowers and fruit in the second year, before withering. Below ground, however, at the start of the third year, they develop a fresh growth of biennial stems that spring up from perennial stems or roots. These grow in regions from the subarctic to the cooler tropical areas across the globe.

Bananas, native to India, Indonesia, and Malaysia, are now grown chiefly in Africa and the Caribbean regions. They are highly nutritious, containing up to 22% carbohydrates and several vitamins and minerals. Some countries rely on bananas as a staple. Pineapples, too, are grown mostly in tropical and subtropical countries. About half the world's export crop is raised on the Hawaiian Islands. Pineapples thrive in a dry atmosphere and in poor, light soil, but require fertilizers and about eighteen months of sun before ripening.

Courtesy French Consulate General

A vintner harvesting wine grapes in Avignon, France.

Coconuts originally grew only on the palm trees of Pacific Island sea beaches, but are now widely cultivated in all tropical regions from India to Jamaica. Date palms are cultivated primarily on the fringes of the deserts of Africa and Asia. These require less water than most trees and can thrive with an underground supply of moisture.

Apples, cherries, and pears are common garden tree fruits usually grown in warm lowlands with good soil, little summer rain, and ample sunshine. Plums, the world's leading stone fruit, are a good example of what can be produced by crossbreeding. Peaches and apricots are native to western Asia, but are now grown throughout southern Europe, America, South Africa, and Australia.

Citrus fruits such as oranges, lemons, limes, and citrons are native to southeastern Asia, but are now grown in many regions with abundant sunshine and frost-free conditions. The grapefruit originated as a hybrid in the West Indies. Tropical trees and bushes which bear fruits such as mangoes, papayas, guavas, and figs frequently ripen and bear fruit throughout the year.

The grapevine, which needs freely draining soil, is grown in every country with a warm sunny climate. A single vine may produce for ten to twenty years. These trailing plants thrive in hilly country with just the right degree of temperature, hours of sunlight, amount of rainfall, and fertility of soil. France and Italy, with perfect conditions for growing grapes, produce half the world's wine.

OTHER CROPS

The largest tobacco crops are produced in Brazil, China, India, Bulgaria, Pakistan, Rhodesia, the United States, and Turkey. As a wild plant, native to the Americas, tobacco is a perennial; in cultivation, it is an annual. It thrives in a considerable range of climates and produces more income per harvested acre than most other crops.

Natural rubber is made from the juice of rubber trees. Nearly all the world's natural rubber comes from southeast Asia, chiefly from Malaysia and Indonesia.

Coffee, tea, and cacao grow in tropical areas. Brazil produces about 40% of the world's coffee, while India and Sri Lanka produce about half the world's tea. Cacao, from which we get cocoa for drinking and chocolate for eating, is an important export of Ghana and Brazil.

Many plants, such as spices and herbs, are raised because they improve the flavor or smell of other foods. Some, such as cinchona for quinine and foxglove for digitalis, are raised for their value as medicines. And still others, such as lavender and certain palm trees, are raised for products like perfume and furniture polish.

Courtesy Malaysian Rubber Bureau

Collecting sap from a rubber tree in Malaysia.

Discussion

1. What percentage of the world's food supply comes from plants?
2. Give some examples of cereals or grains. How are they important nutritionally?
3. Name an important difference in the life cycles of cereals and pasture grasses.
4. What was interesting about the 9000-year-old charred wheat grains found at Jarmo?
5. Name the leading wheat-producing nations.
6. What is the basic dietary staple throughout most of Asia? Of Central America?
7. What is a major use for barley?
8. What special growing conditions pertain to rye?
9. Where is teff the staple grain?
10. What is pastureland used for?
11. What is the average crop yield of an acre of sugar cane? How much sugar can be extracted from it?
12. What are fibers? How are they used? What are the most cultivated fibers? Where are they raised?
13. What are oilseed crops used for?
14. What is vegetative reproduction?
15. Give some examples of root plants and bulb plants.
16. Where do potatoes grow? Why are they important?

17. Name some types of cabbages grown around the world.

18. What does "etioliate" mean?

19. What is a peanut? List some of its uses.

20. Where are bananas grown?

21. Where are most pineapples grown?

22. Give two examples of fruits which are hybrids.

23. Name some tropical fruits that can be harvested throughout the year.

24. Where does natural rubber come from?

25. What are quinine and digitalis?

Review

A. List at least three specific conditions for successfully growing the following crops:

1. Sugar cane ______________________
____
____

2. Cotton ______________________
____
____

3. Potatoes ______________________
____
____

4. Soft-stemmed fruits ______________________
____
____

B. List as many crops as you can think of that grow in a tropical climate, in a sub-tropical climate, and in a temperate zone.

C. What crops are indigenous to your part of the world? to your country? What are the principal crops grown in your country? Is the soil suited for any crops that are not grown there? Does your country produce any major crops not discussed in this unit? Which are consumed in your country? Which are sold to other countries?

D. Name three plants that are annuals. Name three biennials. Name three perennials. What is unusual about soft-stemmed fruits such as strawberries? What is special about tobacco?

UNIT FIVE
FARM TECHNOLOGY

Special Terms

Monoculture: The use of land for growing only one type of crop.
Dryland farming: Growing crops with only natural rainfall as a source of water.
Bund: An embankment built along the side of sloping terrain to halt the erosion of soil.
Milpa: A tract of land which is cleared, usually by burning, farmed for a few seasons, and then abandoned.
Subsistence farming: Producing the minimum necessary to support life.
Plantation farming: A large estate worked by resident laborers, usually growing one basic crop.
Cash crop: Any crop that is easily marketable and is destined to be sold, rather than consumed where it was grown.
Hydroponics: The cultivation of plants in water containing nutrients, rather than in soil.
Truss: A compact cluster of fruit; it is often used to describe tomatoes on the plant.
Reservoir: A natural or artificial place where water is collected and stored for later use.
Water table: The depth below which the ground is saturated with water.

Vocabulary Practice

1. What is *monoculture*?

2. Where does water used in *dryland farming* come from?

3. What is a *bund* used for?

4. What farming pattern is used on a *milpa*?

5. What is *subsistence* farming? *Plantation* farming?

6. What happens to *cash crops*?

7. Describe the process known as *hydroponics*.

8. What is a *truss*?

9. What is the difference between a *reservoir* and a *water table*?

Agricultural Technology

Agricultural technology can be defined as the use of techniques to control the growth and harvesting of plant products, conducted within a framework of vast differences in climate, topography, soil conditions, and social customs. There are more than 3.5 billion acres of arable land surface available for crop farming in the world, but the methods of farming these acres differ according to the local availability of land, labor, and capital.

In addition to rotating crops, fertilizing the soil, and protecting plants from insects, diseases, and weeds, many other agricultural technologies have been developed for specific crops.

Monoculture is not generally successful with non-legume crops because they exhaust the nitrogen in the soil, particularly in humid regions. The availability of low-cost fertilizers has changed this in some areas. It is possible for soil fertility to be adjusted, over a period of time, to fit one crop more precisely than all the crops in a rotation. With several types of soil on one farm, each crop can be grown on the soil best suited to it. Wet areas can be used continuously, for example, for crops not needing early spring tending, while dry soils can be used for drought resistant crops such as sorghum or winter wheat. This type of farming is not widespread, partly because it demands more managerial and technical skills than the crop rotation systems. It also makes the farmer highly dependent on chemicals or cultivation for controlling those insects and diseases that can otherwise be controlled by crop rotation. Moreover, the fertility of the soil must be maintained by application of synthetic fertilizers or livestock manure.

Growing crops without irrigation in regions where less than twenty inches of rain falls each year is called *dryland farming.* The favorite dryland crops are winter wheat, dry beans and peanuts, milo, alfalfa, barley, and cotton. This system is somewhat dependent on rainfall, but knowledge of soil management and nutrient supply is more important. Strips of land are allowed to lie fallow for a season. During this period, moisture is stored for use in the crop season, nitrogen builds up in the soil, and weeds are controlled either by herbicides or by cutting. Modern machinery has reduced the risk of wind and water erosion during the fallow period, but it is important that the proper amount of nitrogen is added during that time through the growth of legumes.

Such great variations in climate exist in the area bounded by the Tropics of Cancer and Capricorn that there is no one system of tropical agriculture. Tropical coasts and other lowlands have a rainy season, which is the principal growing season, and a dry season. During the warm and wet season, the combination of high temperatures and heavy rainfall causes organic matter to decompose quickly, frequently leaving the soil deficient in humus. Mineral nutrients are rapidly released from the soil into plant roots. The soil thus supports the speedy growth of crops, but can be rapidly depleted if not carefully tended. Drainage during this season is also a problem; some minerals may leach or percolate through the soil unless special techniques of water management are practiced. One approach used in Guyana is to build a system of dikes. As a result of this type of controlled irrigation rice and sugar cane production have doubled there.

Soil erosion in the tropics is caused by the intensity, amount, and speed of rain falling on sloping land. The slopes around Darjeeling, India, are so steep that when terraces were built to combat erosion, only a single row of tea could be planted on each one; in Kenya's highlands, it was possible to build broader terraces. Another approach is to build embankments called *bunds*, planted with short grass to stabilize them. Grass strips breed weeds and rodents but they are the favored approach to erosion control in East Africa; a strip three yards wide is left uncultivated along the contour every twenty or thirty yards down the slope. In some places, thatching grass is used to catch the soil while water passes through. Bunds, grass strips, and thatch grass stopwash lines are all effective terraces, if properly maintained.

Courtesy U.S. Department of Interior

Land is sometimes cleared for subsistence farming by burning. Here U.S. Department of Interior workers are using this milpa system to clear sagebrush from land near Twin Falls, Idaho.

Where fertilizer is scarce and land is abundant, farmers frequently clear new lands after depleting the old. As an area of forest is burned, the ashes help to fertilize the land, and weeds are controlled until a new crop can be planted. This *milpa* system, also known as "slash and burn" agriculture, seldom permits more than *subsistence* farming. The principal crops grown for subsistence in the lowlands are maize, sweet potatoes, cassava (manioc), taro, rice, and bananas. At greater tropical elevations, winter imposes problems over and above those caused by seasonal lack of water and soil erosion. The main subsistence crops there are maize, wheat, and rice.

Despite the abundance of plant pests and diseases in the tropical lowlands and highlands, the *plantation* system has flourished there. Specialized production and control practices have developed under skilled management. Plantations are often controlled by companies with direct access to international markets. Tropical plantation *cash crops* include bananas, sugar cane, rubber, rice, coconuts, cacao, spices, palm oil, and pineapple in the lowlands and coffee in the highlands. Specialization in perennial rather than annual crops permits a degree of stability in this hot, wet part of the world.

In some areas of the tropics the major impediment to mechanization is the fear of widespread unemployment. Where low mechanization is uneconomical, new technology has been more rapidly adopted in the areas of improved crop varieties and scientific use of fertilizers. Fertilizers are now less expensive and more readily available than before, and farmers seem to be more aware of their value. The FAO has encouraged the adoption of crop varieties which respond to fertilization, particularly rice and wheat varieties which benefit from nitrogen. Successful use of this new technology requires better credit and fair prices for the small farmer, as well as better distribution and marketing facilities.

Hydroponics, or plant culture without soil, has been technically possible for a long time, but is still expensive on a large scale because

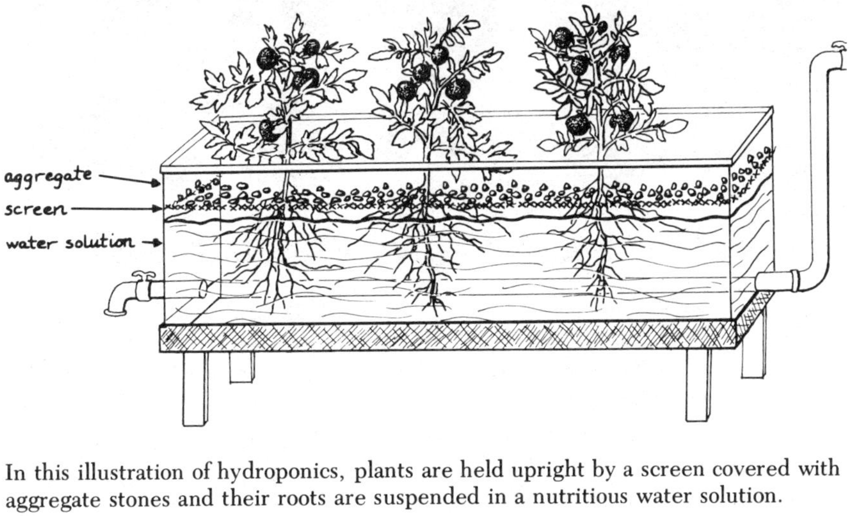

In this illustration of hydroponics, plants are held upright by a screen covered with aggregate stones and their roots are suspended in a nutritious water solution.

it requires the extraction of most of the necessary nutrients from soil. The major advantage of this system is that the grower controls all the foodstuff taken in by the plant and is thereby able to control its life cycle. Hydroponics is usually conducted in greenhouse-type structures and is useful in areas with no soil or extremely infertile soil, such as the Coral Islands of the Pacific.

Greenhouse techniques are being tried for growing ornamental flowers as well as various food crops. In one of the most successful experiments, tomato seeds were planted at a density six times higher than that of conventional methods. The seeds were set in peat and wood fiber, in a shallow tank, into which a nutritious solution was

Courtesy Lord and Burnham

A standard even-span greenhouse.

automatically pumped at regular intervals. The plants were allowed to grow until the first *truss* appeared. The experimenters clipped all further upward growth so that the plant's food was concentrated in the single truss. This method raised the average per acre yield from 45 tons to 134 tons.

Despite this example, greenhouse operations are not widely feasible. The operators must assume many of the tasks normally handled by nature: Regulating temperature and sunshine, providing ventilation, and even carrying out pollination. Mechanization of greenhouse operations has lagged far behind that of agriculture in general.

Irrigation, the controlled, artificial supply of water to land, appeared early on in a simple form: water was taken from the pools left after natural flooding had recessed. Water-lifting devices and dams at the exits of water supplies have been and still are viable systems of irrigation in many parts of the world, but more complicated systems have also been developed. Today, entire rivers are tapped by dams, canals, sub-branches of canals, and trenches of distributary canals. The trench or furrow system is particularly effective for such crops as corn, cotton, sugar beets, and potatoes, where the water can run through furrows plowed between the rows.

Dam and canal systems have profited greatly from modern mechanical engineering technology. Sophisticated regulating devices insure that each canal receives the appropriate amount of water. Modern construction materials help reduce water loss from percolation and evaporation. The larger canals, like those which carry water from Lake Tiberias in the north of Israel to the arid Negev in the south, are made of concrete. Cement-asbestos, rigid plastic, aluminum, and steel are also used, particularly where pipelines are required.

Another technological advance in irrigation systems is the sprinkler method in which mechanical pumps distribute water by an overhead device using a rotating pipe with a series of holes. These are used most often where the land is irregular and difficult to irrigate by the flooding, trench, or furrow systems. Sprinklers are also used where the soil is porous and cannot hold water, thus requiring frequent light waterings. Sprinklers range from those that apply water in the form of a mist to those that apply an inch or more per hour. They are frequently made of lightweight aluminum pipeline and moved so as to distribute water evenly. The cost of such equipment is high, but it has a life span of fifteen to twenty years. Sprinkler system irrigation is used to grow tea and sugar in Africa, coffee in India, and oranges in Central America. In Central America, nutrients are added to water which is sprayed two hundred feet upward by high pressure to help irrigate and fertilize bananas.

Courtesy U.S. Army Corps of Engineers

A U.S. Army Corps of Engineers dam.

Some of the best irrigation systems in the world are those which use canals from large *reservoirs* along the tributaries of the Indus and Ganges Rivers serving Pakistan, India, and Bangladesh. Other countries with highly developed irrigation systems are China, the United States, the Soviet Union, Mexico, the Sudan, and Egypt. The 4,200-mile-long Nile River, with its immense hydroelectric dams, irrigates all the cultivated lands in the Sudan and Egypt. Asia, with the exception of the Soviet Union, contains almost three-fourths of the total of irrigated land in the world.

Modern technology has also helped to solve the drainage problems which arise on all irrigated lands. The *water table* must be controlled and water must move through the soil to wash out salts and bicarbonate deposits which can harm crops. Irrigated fields are often drained by small ditches or tile drains that run into collecting troughs located on lower ground between and parallel to the canals that distribute water. The Netherlands has been able to reclaim land from the sea by using large-scale drainage technology.

All the technological advances discussed here—machines, chemicals, fertilizers, seed varieties, irrigation, and drainage systems—help farmers live better lives and help feed a world that grows by millions daily. But application of new technology is not always easy. Farmers who are illiterate cannot read directions on a bag of fertilizer. Poor roads and transportation can prevent even a high crop yield from getting to market. Limited credit facilities prevent small farmers from taking advantage of modern advances. Competition or poor economic planning can keep prices so low that adoption of a new technology is not profitable. In addition, new methods present risks which traditional farmers may not wish to take.

More technology; more money to finance and apply it; better understanding of existing technology to make it applicable to specific economic, climatic, and cultural conditions, and comprehensive planning are essential to the maintenance of the world's most basic industry.

Discussion

1. Define agricultural technology.

2. Explain the advantages and disadvantages of monoculture. When might this method be effective?

3. What is dryland farming? Why is land left fallow under this system?

4. What is the climate like in the tropics?

5. What happens to organic matter in tropical soil during the rainy season? What happens to inorganic matter at that time?

6. What causes soil erosion in the tropics?

7. Name several methods used to combat soil erosion in tropical lands.

8. What are the principal subsistence crops in the tropics?

9. What are the principal cash crops in the tropics?

10. What are the main advantages and disadvantages of hydroponic farming?

11. Which crops benefit most from the canal/trench/furrow system of irrigation?

12. What are some of the materials used to construct irrigation canals and pipelines?

13. What is a sprinkler? What is it particularly good for?

14. What is unusual about the sprinkler systems used on banana plants in Central America?

15. Name some countries that have highly developed irrigation systems.

16. What has the Netherlands done to increase its supply of arable land?

Review

A. Describe the main systems of irrigation used throughout the world. What system is most widely used in your country? Where does the water come from? What crops are grown with it?

B. Which technology do you think has been most important to your country's agriculture? What are the major impediments to the application of modern technology in your country? What might be done to overcome them?

C. Put the appropriate word or phrase into the following sentences:

1. There are ________ acres of arable land surface in the world.
2. Monoculture makes farmers highly dependent on chemicals to control ________ and ________ .
3. The chief crops grown in dryland farming are: ________ , ________ ________ , ________ , ________ , ________ , and ________ .
4. When organic matter in the soil decomposes quickly, it often leaves the soil deficient in ________ .
5. In Guyana ________ were built to help solve water problems.
6. Grass strips are used to combat ________ in East Africa.
7. The milpa system of farming is also called the ________ method.
8. Some worry that mechanization of agriculture in the tropics will cause widespread ________ .
9. Tropical soil fertilizers are ________ and ________ today than previously.
10. The nutrients used in hydroponic plant cultures come from ________ .
11. ________ of greenhouse-type operations has greatly lagged behind the pattern of agriculture in general.
12. A ________ is used to block the natural exit of a water supply. ________ then carry the water to fields to be irrigated.

13. Sprinklers are often made of ________ .
14. Three-fourths of the irrigated land of the world is in ________ .
15. Drainage ditches often run ________ and ________ to water distributing canals, but on lower ground.
16. The application of techniques to control the growth and harvesting of plant products can be called ________ .

UNIT SIX
ANIMAL AGRICULTURE

Special Terms

Fowls: Several species of birds important as sources of meat and eggs. Chickens, turkeys, geese and ducks are examples.

Ruminant: An animal with a complex, multi-compartment digestive system which allows it to utilize coarse feedstuff.

Rumen: The first and largest digestive compartment of hoofed animals like sheep, in which bacterial action breaks down coarse materials and synthesizes amino acids and B-vitamins.

Amino acids: A group of organic acid compounds which combine to form various types of proteins.

Forage: Coarse feed materials high in fiber and water but low in energy.

Pasture: Land on which forage is grown and consumed directly by animals.

Hay: Forage which is cut, dried and stored for use when growing feed is not available.

Silage: Crops which are cut while they are green and stored as animal feed.

Ration: Diet of a farm animal.

Concentrates: Feeds which are high in energy and low in water, such as grains.

Molasses: A thick, sweet liquid high in sugar content, obtained by boiling crops such as sugar cane, sorghum, and corn.

Confinement: A method of raising livestock in a small area where environmental conditions can be controlled.

Breed: A class of related animals with similar traits, such as color, presence of horns or facial characteristics, which can be passed on to the offspring with little variation.

Artificial insemination: Introduction of semen into a female animal by mechanical means.

Purebred: An animal descended from ancestors of the same breed.

Quarantine: A period of enforced isolation during which animals are monitored for signs of contagious diseases.

Vaccine: A serum containing specific micro-organisms administered to increase immunity to certain diseases.

Parasite: An organism which lives from, on or within another organism.

Vocabulary Practice

1. Give some examples of birds which are *fowls*.
2. What special ability does a *ruminant* animal have?
3. What substances are synthesized in the *rumen*?
4. What are *amino acids* and what do they form?
5. What are the nutritional characteristics of *forage*?
6. How is *pasture* used by livestock farmers?
7. How is *hay* produced and for what is it used?
8. How do *hay* and *silage* differ?
9. What does a *ration* refer to?
10. What nutritional characteristics do *concentrates* have?
11. How is *molasses* obtained? What does it contain?

12. Why are some animals raised in *confinement* buildings?

13. What do animals of the same *breed* have in common?

14. What is the meaning of *artificial insemination*?

15. How does a particular animal qualify to be a *purebred*?

16. Why are some animals subjected to a *quarantine* period?

17. What is a *vaccine*? Why is it used?

18. How do *parasites* live?

Animal Agriculture

People learned to utilize animals very early in history. Many primitive tribes hunted wild animals for food. They also learned to make clothing and shelter from animal hides, and used the bones and tusks to make tools, weapons, and ornaments. The importance of animals to the survival of early humans is borne out by the appearance of many animal-like figures among the deities worshipped by ancient civilizations.

As centuries passed, various civilizations realized that hunting was an undependable means of obtaining food. Animal herds roamed over vast territories, making game sparse. Drought or disease could sharply reduce the population of some species. Hunting became dangerous if the animal being chased decided to attack its pursuers.

Domestication of animals began very early in certain parts of the world. Cattle raising began in southwestern Asia by 5000 B.C. Although most often utilized for meat, cattle also provided milk and hides, and served as beasts of burden. Horses were domesticated by 3500 B.C. in the Persian Gulf area, and were used mostly as means of transportation. Later important species included sheep, swine, goats and *fowls*.

The most common use of domestic animals has been as a source of food. Cattle and goats provide both meat and milk. The milk is further refined into such products as cheese, butter, ice cream, and yoghurt. Sheep have also provided meat for many centuries. In the Western Hemisphere, Europe, and China, swine contribute a major portion of the meat supply. Poultry such as chickens, turkeys, ducks, and geese, once delicacies, have become a major meat source in many countries. They serve a dual function as producers of eggs as well as meat.

Animal hides have long provided people with protection from the extremes of nature. Indian tribes of the North American plains used buffalo and deer hides for shelter. Wool from sheep, goats, and animals like the alpaca and vicuña of South America has been a raw material for clothing for thousands of years. Even today fine leather goods are much in demand.

Some species of animals became beasts of burden, and are still used as such in many parts of the world. Horses, donkeys, camels, llamas and even elephants were carrying heavy loads long before trucks and trains appeared. Most of these species also serve as means of human transportation.

Farmers also found that certain animals were well-suited to pulling implements which prepared the soil for planting and harvesting crops. Horses and cattle have been the most popular work animals. The water buffalo is a particular type of cattle commonly used as a source of power in Asia. In North America, livestock breeders found that the mating of a horse with a donkey produced a hybrid called a mule, an animal which was exceptionally well-suited to pulling farm implements or carrying heavy loads. A good team of oxen or mules may be a small farmer's single most important investment.

RUMINANT ANIMALS

Cattle, sheep, and goats enjoy a particular distinction among domesticated animals because they are *ruminants*. A ruminant has a complex digestive system which includes four "stomachs." The largest of these stomachs is the *rumen*, a shapeless compartment which provides an environment for many thousands of tiny bacteria. These bacteria break down the tough cellulose fibers of coarse materials eaten by the animals into substances which can be more easily digested by the other simpler stomachs.

The rumen also synthesizes important *amino acids* and B-vitamins. Amino acids are protein building blocks used in forming meat, bones, blood, hair and skin. Some varieties of amino acids are essential for building tissues in animals and humans. Ruminants have the ability to synthesize certain of these essential amino acids from others contained in the feeds they eat. Simple-stomached species, such as swine and humans, must obtain all the essential amino acids from their diets.

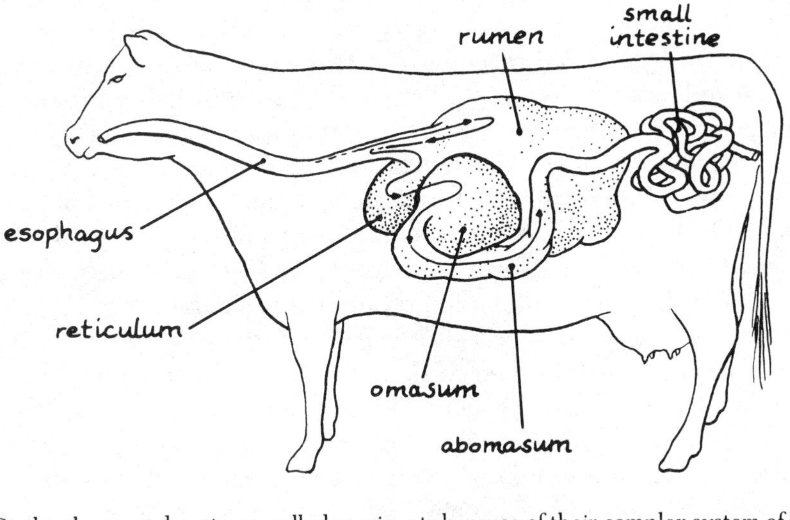

Cattle, sheep, and goats are called ruminants because of their complex system of four "stomachs." This illustration shows the path which food takes as it is digested.

The ruminant animal's special abilities allow it to eat types of food which humans cannot consume directly. The most important of these are grasses, legumes and other types of forage high in fiber and water but low in concentrated energy and protein, such as grains. Often such forage is grown on land that is too steep, too dry, too wet or too infertile to grow crops for human consumption. Thus cattle and sheep can add meat and milk to human diets without directly competing for other food producing resources. They have the added ability to consume forage directly through grazing, eliminating the need to expend human labor in harvesting and transporting feed.

Where grazing land, or *pasture,* is abundant, cattle herds and sheep flocks may be allowed to range over great distances, sometimes accompanied by one or two herders to watch over them. Where land is more limited, intensive grazing may be carried out. Highly productive and specially adopted varieties of grasses and legumes are seeded. Pastures are fenced and livestock rotated to control depletion. Fertilizers and pesticides may be applied to increase pasture production.

In some fortunate areas of the world, pasture is available throughout the year. In other regions, however, low rainfall or cold temperatures during certain seasons make it necessary to provide livestock with supplemental feed. Forage may be dried and stored in the form of *hay*. Sometimes forage and grain crops such as corn are cut while they are green and stored as *silage*. Care must be taken to prevent fresh air from reaching the silage and causing it to spoil. Specialized storage towers (called silos), trenches in the ground or even giant plastic bags are used to preserve silage. Corn, alfalfa, sugar cane, sugar beets, milo, and sudan grass are common silage crops.

In areas where grains such as corn, soybeans, wheat, and milo are abundant, they, too, are important sources of feed for cattle and sheep. Cattle can be grown to maturity and fattened on a strictly forage diet, but such a diet often requires three to four years. On the other hand, high energy *rations* containing grains or other *concentrates* may allow cattle to reach slaughter weight in as little as fifteen months. There is some tendency for the younger, fatter animals to produce more tender and juicy meat, although age is probably less important than the way in which the meat is aged and cooked.

Milk production is also affected by the amount of energy consumed by the cow or goat. In many places milk is produced profitably from pasture and other forage alone. However, the addition of concentrated energy sources such as grain or *molasses* has been shown to significantly increase the quantities of milk and butterfat produced, particularly in cows with the genetic potential for high production.

Whether cattle and sheep are fed primarily forage or concentrates depends on the supply and demand for each type of feed in the area. Where grain production is low and is utilized mostly for human consumption, livestock feed must come from land not suited to production of more valuable crops. In grain surplus regions farmers will combine forage and concentrates in the proportions resulting in the lowest overall cost and most rapid growth.

The practice of feeding grain to livestock has been criticized as an inefficient use of potential food supplies. History has shown that grain does shift from livestock feed to human consumption when worldwide demand for food is strong. As grain prices rise, farmers sell their grain and feed more forage to livestock, or reduce livestock production

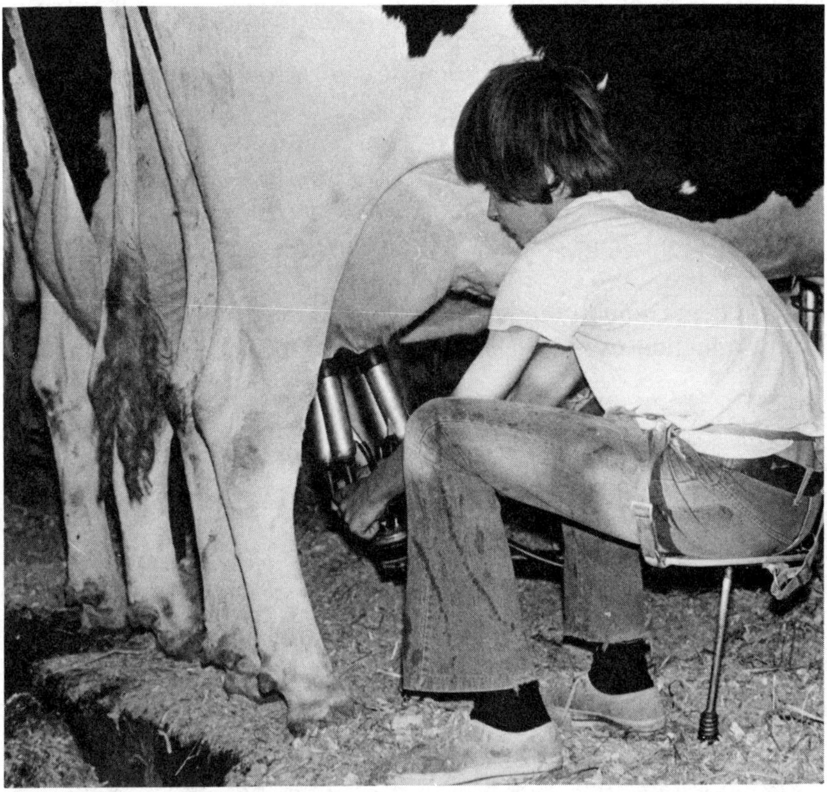

Jerome Wexler/Photo Researchers, Inc.

Here a farmhand milks a cow using a milking machine.

altogether. But the nutritional needs of people around the world must be translated into an economic demand strong enough to both compete with the demand for livestock products and pay the considerable costs of transporting grain internationally before such shifts in grain usage occur. This means that purchases from grain surplus nations must be made either by the grain deficit nations themselves, or by charitable groups, international agencies or the governments of other grain surplus countries.

NONRUMINANTS

Nonruminant or simple-stomached animals include swine, chickens, and other types of poultry. Their digestive systems are similar to those of humans, and they must often compete for the same food supplies. Swine and poultry are efficient converters of feed into meat, however. In addition they are prolific. A hen may produce over two hundred eggs per year, while a female pig (or sow) produces from ten to thirty offspring in a year.

Both pigs and chickens have been common "backyard" enterprises for centuries. They require very little space, and can consume scraps from the table and garden. However, both species are susceptible to a number of diseases. Drugs are now available to combat many of these.

A recent trend in methods of raising swine and poultry has been the introduction of *confinement* buildings, special purpose housing

George Laycock/Photo Researchers, Inc.

With laying hens in confinement, one man can tend thousands of birds. Water is supplied automatically and eggs roll down the sloped cage floors to be removed for packaging and distribution.

developed to control the environment of the animals. Temperature, humidity and ventilation are kept at near optimum comfort levels. Placing a large number of animals in a small space also reduces the amount of labor required for care and feeding. Just one person can look after a thousand head of hogs or several thousand birds. Automatized removal of excrement or manure is more easily accomplished when animals are confined, but safe, odorless disposal of manure still presents challenges to both farmers and engineers.

ANIMAL BREEDING

The science of genetics has contributed much to modern livestock production. For centuries the processes of natural selection and survival of the fittest helped develop animal strains with resistance to heat, cold, insects, disease, drought and other adverse environmental elements. While this enhanced the animals' ability to survive, it did not guarantee that other desirable traits would be present.

Scientists first used genetic principles to influence the characteristics of certain plants. Later geneticists and livestock producers turned their attention to systematically improving such animal characteristics as muscling, milk production, prolificacy, growth rate, and size. Other traits such as wool quality, absence of horns and coloration have also been controlled through selective breeding.

In many isolated areas, locally grown animals had been inbred, that is, used as parent stock without the introduction of unrelated animals, for many years. The offspring all had very homogeneous traits. This is how various *breeds* of livestock originated, and why two parents of the same breed can be expected to produce offspring nearly identical to themselves. In some cases offspring with superior qualities resulted. By selecting these offspring for breeding stock, the overall quality of the breed was raised.

Many experiments were done in which crossbreeding (crossing two animals of different breeds) resulted in offspring with desirable qualities from each parent. For example, crossing British origin beef breeds with Brahma cattle from India produced meat-type animals who could tolerate the heat and insects found in tropical climates.

In recent years many large, fast growing beef breeds from Europe have been introduced into the United States and other countries for crossing with the smaller, more compact traditional British breeds. Their offspring grow faster, require less feed per pound of meat, and produce a carcass with less fat. Hog producers are also trying to reduce the amount of fat found on swine carcasses. Animal fats used to be valued for making soap and candles, and as a concentrated energy source, but this is no longer true in most of the world.

The development of *artificial insemination* has allowed certain superior sires to pass on their genetic potential to many offspring. Some outstanding bulls have fathered more than 100,000 sons and daughters. Frozen semen can be kept for many years, even after the donor animal has died, and has been responsible for rapid genetic progress, particularly among dairy animals.

Many livestock breeders maintain offices where *purebred* animals can be registered, that is, have their pedigrees recorded. Performance records of close relatives are also available. This allows buyers to gauge the potential of young breeding stock for passing on desirable traits.

ANIMAL HEALTH

Most domestic livestock are susceptible to the same multitude of diseases and ailments as humans. Moreover, they cannot as easily express their discomfort or their symptoms. Where certain species of livestock are found in concentrated numbers, diseases can spread rapidly. The dreaded foot-and-mouth disease spread across Mexico in 1947-1954, and was elimiated only by isolating and slaughtering all infected cattle. More recently African swine fever has spread through the Caribbean countries. In some countries such as Haiti and the Dominican Republic, complete destruction of all hogs has been the only solution to controlling it.

Most countries now place restrictions on the importation of livestock. Often a *quarantine* period is required before animals are released. Many countries have passed strict laws requiring that animals infected with certain highly contagious diseases be destroyed. The government may repay the affected owner for his losses.

Vaccines have been developed for many serious livestock diseases. Newcastle disease in poultry, cholera in cattle and hogs, and scropie in sheep have largely been controlled in the United States as a result of widespread vaccination programs.

Some diseases are transmitted by insects and can be controlled by sanitary methods. Dipping or spraying livestock with insecticides, removing manure promptly, sterilizing livestock facilities before introducing new animals and rotating livestock among different locations all help to break the life cycles of the disease-carrying organisms. *Parasites* such as intestinal worms and fly larvae, which cause slow growth or even death, can be controlled by the same methods.

Disease and insect control is particularly vital in temperate climates where freezing temperatures which kill many dangerous organisms each year do not occur.

FUTURE OF ANIMAL AGRICULTURE

The strong demand for animals and animal products continues as people in developing areas of the world press for improved levels of nutrition, particularly protein consumption. This demand for livestock as a food source will probably more than offset the declining use of animals as sources of power and transportation.

Scientists working in the area of livestock production face many challenges, perhaps the most important of which is finding suitable feed sources which do not compete with resources used for human food production. New sources of protein and energy, improved methods of processing feed to increase its nutritional value and the use of growth stimulants promise to improve the efficiency of livestock production.

Animal health specialists must learn how devastating livestock diseases are spread and how to control them. As international trade and travel increase, the potential for worldwide epidemics grows. Increased distribution of superior breeding stock should improve the quality of animals available to farmers.

Increased funding of research and exchange of information can make these advances possible. As a result farmers will be able to produce livestock more efficiently and profitably, and consumers will benefit from a more complete and abundant food supply.

Discussion

A. What kinds of animals are important to agriculture in your country? What is each species used for? Are some species becoming more important? Less important?

B. What are beef and dairy cattle fed in your country? Does it vary according to region? Is feed more abundant during some seasons than others? What can be done to alleviate the problem of seasonal feed shortages?

C. Do you think it is wrong to feed grain to livestock when malnourishment exists in the world? Why is grain fed to animals instead of people? How could this be changed? Can livestock actually increase the total food supply in the world?

D. What areas of science have livestock growers used to improve their production? Who benefits from new discoveries in livestock production practices?

Review

1. For what purposes did primitive civilizations use animals? What other uses came about later in history?
2. What advantages did domestication of animals have over hunting wild animals?
3. In what areas of the world were animals first domesticated?
4. Name four species of animals used as a source of meat.
5. What other food products can be obtained from animals besides meat?
6. Name three species of animals used to transport humans or other heavy loads.

7. What other functions do animals perform besides providing food and transportation?

8. What type of animal used for pulling is a hybrid of two other species? What are the two other species?

9. What advantage does a ruminant animal have over a nonruminant for survival? Give examples of ruminants.

10. How does the rumen digest coarse food?

11. Where do simple-stomached animals obtain essential amino acids?

12. What can be done to increase pasture production where land for grazing is limited?

13. Why can't livestock graze throughout the whole year in some parts of the world?

14. What crops are used to make silage? How can silage be preserved?

15. What differences result in livestock production if cattle are fed a ration high in concentrates rather than forage?

16. Give some examples of nonruminant animals.

17. What are the advantages and disadvantages of raising livestock in confinement buildings?

18. What traits did animals develop as a result of natural selection? As a result of scientific breeding?

19. Give an example of improvement in livestock production achieved through crossbreeding.

20. How has artificial insemination helped to improve livestock quality?

21. What measures can countries take to prevent and control livestock diseases? What can individual farmers do?

22. How are insects harmful to livestock?

UNIT SEVEN
THE BUSINESS OF FARMING

Special Terms

Agribusiness: A complex of agricultural activities including farm production, manufacture and distribution of farm supplies and equipment, and processing, storage and distribution of farm products.

Cooperative: An enterprise owned collectively and operated for the mutual benefit of the participants.

Land reform: Programs undertaken by national governments to redistribute large agricultural holdings.

Nomads: Those who have no fixed residence, but move from place to place in search of pasturage or food.

Processing: The conversion of agricultural commodities into marketable or usable forms.

Marketing: Moving commodities from producers or processers to consumers.

Grade: A product classification or standard based on quality, size, color, or judgment.

Food additives: Substances introduced into food for the purpose of changing it in some way.

Food preservation: Any of a number of ways of retarding natural food spoilage.

Freezing: Storing food at low temperatures to prevent the growth of micro-organisms.

Pasteurization: Heating milk or other beverages to a temperature high enough to kill organisms which could cause spoilage.
Blanching: Scalding or steaming food to remove skins or retard spoilage.
Irradiation: Exposure of food to light or other radiation sources to prevent germination from occurring.

Vocabulary Practice

1. Describe the activities of an *agribusiness*.
2. What is a *cooperative*?
3. What does the phrase *land reform* refer to? Can you think of any country that has had such a program?
4. Use the word *nomad* in a sentence referring to agriculture.
5. Define *processing*.
6. What function does *marketing* perform?
7. What determines a *grade*?
8. What are *food additives*? Are they basic parts of foods?
9. What does *food preservation* mean?
10. Describe how these processes help to preserve food: *pasteurization*, *blanching*, *irradiation*, and *freezing*.

The Business of Farming

To say that the food industry, with farming itself at the core, is the largest and most important in the world is no overstatement. Farming ranks with the manufacturing, construction, transportation, and service industries as a major component in the economy of the industrialized nations of the world. The United States, for example, with less than 7% of its labor force engaged in farming, is by far the largest exporter of farm products in the world. Other major exporters of farm products include Canada, Australia, New Zealand, Argentina, and the Soviet Union.

In less developed countries, there is generally less efficient use of labor and smaller farms than in the more developed nations. In these cases the agricultural industry, which includes services supplied to farmers and marketers, cannot frequently obtain or afford new types of technology. Many farmers in these countries produce food chiefly for their own families, selling or trading the leftover produce. Others work as laborers for those large operations whose business is the growing, processing, and marketing of foodstuff—*agribusiness*.

Approaches to organizing this industry vary widely throughout the world, depending on the type of agriculture, the philosophy and practice of government, the stage of economic development, and the societal customs and traditions. In some countries, the typical farm is a family-run private enterprise organized to produce profit. The land, machines, buildings, materials, and labor are rented or owned by the family. Capital is owned or borrowed for investment. In other countries, the farm resources are organized into *cooperatives* or are owned and operated by the state. In most countries, there is a broad mixture of varying degrees of individual ownership, cooperative enterprise, and governmental control. In parts of Latin America, despite a trend toward *land reform* and the existence of numerous small farms and cooperatives, the largest proportion of farm resources is concentrated in large, privately owned holdings. In most of tropical Africa, where the FAO estimates that up to 80% of the population is engaged in agriculture, there are literally millions of small farms. All types of crops and farms exist in Asia, from the *nomad* agriculture of the Arabian desert to the highly organized, traditional, peasant, state-controlled economies of China and India.

No matter what type of organization, degree of sophistication, or class of crops grown, all farms have some problems in common: They require management, they must have financing, and they must market their produce.

Successful farm management requires knowledge of crops, machinery, cultural practices, accounting, marketing, and purchasing. Organization of labor, knowledge of farm technology, and adherence to economic principles are vital. Farm management research is particularly needed in developing areas. New high-yield cereal varieties, for example, can achieve full potential only if they are cultivated with adequate inputs of fertilizer and water, careful attention to crop protection, and high standards of planting procedures. Proper water management and development of water resources are particularly important in regions like the Asian subcontinent where rainfall is concentrated into a few months of the year.

Labor-efficient farming requires heavy capital investment. As greater output per worker is achieved, the amount of money needed per farm, or per farm worker, increases dramatically. In many parts of the world farms are growing larger, the number of farms is decreasing, and the debt per farm is increasing. In these areas there is relatively reasonable access to credit from banks, agricultural lending agencies, and merchants. In other areas, where the supply of credit is insufficient, there are often additional problems which hamper agricultural development as well. The supporting or input industries involving storage, warehousing, transportation, and farm utilities are frequently inadequate. Better crop yields alone do not insure sufficient food supplies.

Large scale marketing of farm produce and greater specialization in product *processing* and handling are more evident in a well developed economy. Such inputs as fertilizers, insecticides, managerial planning, and consulting services from all parts of the food industry aid the farmer's efficiency. Delivery of such products and services from a wide spectrum of the food/agricultural industry benefits the farmer directly. *Marketing* firms may offer specialized services; some even contract to buy farm products by prior agreement. Since the marketers then have a direct interest in high yield and good quality, they offer such services as *grading*, packing, and drying of products.

In less organized or developed areas, much of what is grown cannot be shipped long distances and is inadequately stored. And

Courtesy Farmland Industries

Packaging of agricultural products serves the dual purpose of advertising and protection of the product. Here bacon is hermetically sealed in packages that display the product.

despite the limited markets, the measures of volume, quality, and price are not uniform. Therefore, quality is hard to evaluate and maintain.

Packaging plays an important role in resolving some of the marketing and distribution problems shared by the world's farmers. While some aspects of packaging are aimed at simple identification and the promotion of sales, many types of packaging help protect the appearance, taste and nutritional value of the products. Preparation of goods for distribution and sale involves the use of glass and plastic bottles, metal cans, cardboard cartons and boxes, waxed paper containers, plastic and paper bags, metal foils, drums, and many other items. Two of the most interesting aspects of the processing and packaging activities carried on by the food industry involve *food additives* and *food preservation*, usually regulated by governmental agencies. Indeed, because of the potential impact on the health and financial well-being of the consumer, the food industry as a whole is one of the most closely regulated in the world.

Courtesy USDA

A U.S. Department of Agriculture meat inspector in Omaha, Nebraska, is shown at work in one of more than 3,000 federally inspected meat plants in the U.S.

The use of food additives dates back to the salt, smoke, vinegar, and spices used in antiquity. Today's additives are much more sophisticated, however; acids protect against decomposition or fermentation, synthetic flavoring agents impart flavor and adjust for seasonal variations, food coloring changes appearance. Some additives retain or improve texture, aid in processing or manufacture, and facilitate packaging. Perhaps the most important additives are those that add vitamins and minerals or replace those which have been lost due to some other process, such as the milling of grain or the pasteurization of milk.

The earliest methods of food preservation were smoking, drying, salting, fermenting, and pickling. Today there are several basic modern methods used for preserving foods. One uses heat to kill or suppress the growth of microorganisms and to activate enzymes. Canned food is heated and then sealed in airtight containers to prevent contamination. *Freezing* suppresses microbial growth and hinders biochemical changes. The chief preservatives in use today include dehydrates, antioxidents and antisprouting agents, and updated versions of the earliest methods in which salts, sugars, acids, smoke, and other natural agents kept food edible indefinitely. Some foods require specific techniques, such as *pasteurization* of dairy products, *blanching* of certain fruits, and *irradiation* of potatoes.

Government agencies and private industry are constantly striving to efficiently deliver farm products of high and uniform quality to consumers at minimum cost. Scientists, engineers, and those in all phases of the farming business conduct research on handling, processing, sorting, labeling, packaging, storing, transporting, and distributing agricultural products. Not the least of these areas of research is aimed at improving means of transportation. Without adequate transportation facilities we would all be limited to the foods that grow in the climate and soil of our own neighborhoods. There is no one solution to the transport problem; consumer demand affects the development of some devices for fast, effective delivery.* On balance, transport equipment and services rank high as an area in which improvement is needed. To hold down transportation and handling costs, and to reduce spoilage losses, it is necessary to improve present loading, refrigeration, and cargo securing systems for shipping agricultural products.

Courtesy Farmland Industries

The fast and efficient transport of products from the farm to the store is an aspect of the agricultural distribution system that can be improved upon.

*A cost/time of delivery comparison for one ton of animal feed going from Los Angeles to New York (3000 miles): air—$400/6 hours; truck—$90/5 days (on good roads); train—$35/7 days (on fast lines); ship—$30/15 days (through the Panama Canal in good weather). How quickly is the product needed? How much is the consumer willing to pay for speed? Many factors affect the transportation decision.

Agriculture in any given nation depends on and is determined by climate, soil, population density, availability of arable land, and use of varied technologies, as well as national and international economic priorities. Britain, for example, could be self-supporting in dairy products, but needs markets for its industrial products. As a result, the government has designed incentive systems and sanctions so that fewer farmers engage in dairy farming. The British now import several hundred million dollars worth of dairy products each year from countries which in turn buy industrial products from them.

The business of farming is complex enough when viewed separately. It is further complicated by the fact that farmers must rely on the petroleum industry for fuel, on the international balance of trade for prices they receive, on global marketing and transportation systems, and on myriad other factors. Weather and soil are important, but so are politics, economics, and human desires.

Discussion

1. How does the food industry compare in importance to the other major industries of the world?

2. Which countries are the world's leading exporters of farm products?

3. Name some of the factors which influence how a country's agricultural resources are organized.

4. What are some characteristics of a family farm?

5. How are farm resources often organized in Latin America?

6. How much of the population is engaged in farming in tropical Africa?

7. What types of problems do farmers around the world have in common?

8. What knowledge is necessary for successful farm management?

9. What trends are noticeable in farms as a country becomes more developed economically?

10. Give an example of a marketing firm's involvement with farm resources. How does this benefit the farmer?

11. What are some of the problems encountered by farmers in poorly developed economies?

12. Why are foods packaged?

13. What are some of the materials used to package foods?

14. Why is regulation of the food industry necessary?

15. What were the earliest food additives?

16. What are some of the things modern additives do?

17. What is the most important type of food additive used today?

18. What were the earliest methods of food preservation?

19. Give some examples of food preservation used today.

20. Why is transportation important to the food industry?

21. What are some of the problems encountered in transporting food from the farm to the consumer?

22. Why has Britain decided not to be self-sufficient in dairy products?

23. Name some of the factors that influence the farming business.

Review

A. Match the short definitions in Column A with the words or phrases in Column B:

Column A	Column B	
1. form of transport of farm products	plastic bottle	________
2. collectively owned and operated farming enterprise	food coloring	________
3. packaging form	decomposition	________
4. food additive	cooperative	________
5. food preservation method	farm credit	________
6. type of food spoilage	marketing	________
7. type of food processing	trucking	________
8. sale and distribution of agricultural products	freezing	________
9. source of capital for investment	sorting	________

B. Describe the skills necessary to a successful modern farmer. Which of these skills do you now possess? How can you acquire the others?

C. How are farm resources organized in your country? Think about the relative merits of the private enterprise system and the various state controlled or owned systems. Which do you think would work best in your country? Why?

D. How available is farm credit in your country? Improved seed? Synthetic fertilizers?

E. The next time you buy food, think of these things: How was the food grown? Where? Was it processed? How? What packaging materials have been used? Are there any food additives? If so, why? Why did you choose these particular items?

UNIT EIGHT
THE FUTURE OF AGRICULTURE

Special Terms

Eutrophication: The accumulation of so many mineral and organic nutrients in freshwater lakes and streams that the oxygen is depleted; this environment favors plant life but is destructive to animals and fish.

Algae-laden swamps: Wet, spongy, marsh lands with dense vegetation, unfit for cultivation.

Environment: The collective name for all the external conditions and influences affecting the life and development of the earth.

DDT: A colorless, water-insoluable insecticide, toxic to people and animals when swallowed or absorbed through the skin. The letters are an abbreviation of the chemical name dichloro-diphenyl-trichloroethane.

Biosphere: The part of the earth's crust, waters, and atmosphere in which life can exist.

Marginal: Land with the minimal requirements for crop production.

Deforestation: Complete removal of tree cover from the land.

Particulates: Extremely small particles of matter which do not naturally occur in the air, but are byproducts of wind and soil erosion.

Atmosphere: The gaseous envelope of air surrounding the earth; part of the biosphere.

Laterization: The process of soil solidification. In certain types of tropical soils, after the nutrients are leached an infertile, red or yellow substance rich in iron oxide, remains. This converts quickly to a bricklike rock called *laterite*, which was used to build the temples at Angkor Wat.

Plankton: Plant and animal organisms that drift in great numbers in water.

Green revolution: The development of high-yielding varieties of wheat, rice, and corn and associated with fertilization and cultivation practices. The wheats were developed first in Mexico by the Rockefeller Foundation under the direction of two Americans, George Harrar and Norman Borlaug. Borlaug won the Nobel Peace Prize in 1970 for this work.

Desalinization: The process of removing salt from seawater.

cwt: Hundredweight. A unit of weight equal to 100 pounds in the United States, 112 pounds in Britain, and 45.36 kilograms in the metric system.

Reclamation: The process of making waste, desert, marshy, or submerged land arable.

Vocabulary Practice

1. Describe the process known as *eutrophication*.
2. What is an *algae-laden swamp*?
3. What does the *environment* refer to?
4. What is *DDT* used for? What do the letters mean?
5. What is the *biosphere*?
6. What is *marginal* land?
7. What does *deforestation* involve?
8. How do *particulates* get into the air? What are they?
9. What is the gaseous envelope surrounding the earth called?
10. Describe the process that transforms soil to *laterite*.
11. Where is *plankton* found?

12. Give a brief description of the green revolution.

13. What is the purpose of *desalinization*?

14. What is *cwt* equal to in the metric system?

15. What is the process of making wasteland arable called?

The Future of Agriculture

Few problems facing the world today are more acute than those involving production and distribution of food for our fast-growing population. United Nations projections indicate a near doubling of the earth's population by the year 2000. Feeding these numbers adequately would require more than a doubling of current food production since, at present, a high proportion of people are chronically undernourished. Recently a Honduran nutrition expert estimated that between 75% and 95% of the children in that country suffer from some form of malnourishment. Our food-producing capacity is less than adequate now, and the pressure to meet future demands will be enormous.

While there are signs that the problems of increasing our food supply can be solved, other problems have arisen as a result of our attempts to do so. Among these is the *eutrophication* of freshwater lakes and streams. The use of fertilizers all over the world has greatly expanded since the introduction of new cereal strains which require enormous amounts of added nutrients. These nutrients, in turn, frequently wash into freshwater bodies, depriving fish of oxygen and turning these lakes into *algae-laden swamps.* In many parts of the world fish are an important part of the diet. Eutrophication is even reducing the fish supply in the oceans. In effect, the modernization of agriculture increases some food supplies while it sacrifices others.

Another major area of concern resulting from agricultural intervention in the *environment* is the endangerment and loss of whole species. The two main causes of this alarming phenomenon are the reduction of natural habitats and the introduction of toxic elements. As land is cleared for cultivation, the sources of subsistence diminish for animals. For example, wild elephants in Sri Lanka and

Bengal tigers in India are greatly threatened by the loss of wilderness area. In this century, an average of one species per year has become extinct somewhere in the world, and the number of endangered mammals and birds is in the hundreds. The peregrine falcon, the American bald eagle, and the Russian wild duck all face extinction from pesticides introduced into the environment. The virtual extinction of storks that winter in North Africa and summer in Denmark is due to pesticides which protect crops from locusts in the Nile Valley. Some of these processes are already irreversible and are complicated by severe problems of human hunger. The use of *DDT*, for example, has been banned by many governments because of its adverse effects on the *biosphere* and the resulting harm to humans and animals. DDT, however, is used not only for agricultural purposes,

A World Health Organization helicopter sprays DDT on foliage in Kinshasa, Zaire. The 1961 aerial assault was to stop a plague of malaria mosquitoes.

Courtesy Argonne National Laboratory

The land in the foreground is badly scarred by erosion; the fields in the background are not eroded.

but for killing the malaria-carrying mosquitoes that plague tropical and subtropical regions. Many countries whose concerns are feeding people and eradicating malaria—rather than saving birds—have no intention of banning its use.

Soil erosion and climate modification are two more by-products of problems caused by modern agricultural activities. Much of the land cultivated for the first time is *marginal;* the topsoil is thin and quickly loses its arable value. The need for more grazing land for livestock also strips the land of its natural grass cover.

The land is then ripe for erosion by wind and water. The natural cover that holds soil and retards water runoff is reduced, and floods and drought often follow. The Fertile Crescent once supported large human and animal populations but is now arid, largely because of overgrazing and soil abuse. Soil moves naturally toward the ocean, but when agricultural activities remove portions of the vegetation cover, that movement accelerates; the topsoil is thinned to a point where it is no longer productive, at least under current technologies. Each year literally millions of acres of cropland in North Africa, Asia, the Middle East, and Central America are abandoned because soil erosion by wind and water has made the land incapable of sustaining the local inhabitants. In parts of Spain, Greece, and Korea, countrysides once covered with forests and grasslands are now dust bowls. *Deforestation* in nothern Brazil has caused droughts and resulting famine in that region. Today's water crisis in Mexico is rooted in the deforestation of parts of that country three hundred years ago. This severe problem itself has a by-product—an increase in *particulate* matter in the *atmosphere*. There is evidence that this increase is reducing the average yearly temperature, reducing the amount of solar energy reaching the earth, and causing a southward shift in the frost and ice boundaries in the northern Atlantic Ocean. Such climatic changes, carried to extremes, could ultimately produce snowfalls which will not melt during growing seasons in the temperate zones.

Another problem associated with expansion into untapped tropical soils is *laterization*, or soil solidification. This process is thought to have undermined the Khmer civilization in Cambodia centuries ago; it has destroyed the Brazilian experimental station at Iata in recent years.

What positive signs are there for the future of agriculture? Some of the advances in the fields of energy, equipment and machinery, science, technology, and processing have been discussed here. Foods have, indeed, become more plentiful and more varied as a result of science and technology. Farm machinery makes it possible to grow more food with less labor, and modern processing has increased the variety of foods. New fruits and vegetables have been crossbred. Advances in animal feeding have increased their yield of meat. Even cooking has been made easier by such methods as quick freezing, dehydration, and microwave heating. One promise of future nutrition supplies comes from current experimentation with *plankton*, the microscopic plants and animals that abound in the oceans.

The so-called *green revolution* has created new high-yield cereal crops which are unusually responsive to chemical fertilizers and water. Some of these wheat and rice varieties have already had a profound impact on Indian and Pakistani agriculture. In some areas, an eight-fold increase in crop production per acre has been achieved. The Philippines, home of the International Rice Research Institute, is self-sufficient in rice for the first time in fifty years. The amount of corn and wheat raised per acre in Mexico has more than tripled in the last fifteen years. These new strains have revolutionized agriculture in many parts of the world. Not only can more grain be grown on less land, but the product is more nutritious and is adaptable to almost every type of soil in the tropics and subtropics.

New sources of protein are being exploited in some areas, offering hope to the poor and undernourished. Vitasoy is a soybean-based beverage now being produced in Hong Kong; it is high in protein and constitutes about 25% of the soft drink market there. Saci, a Brazilian beverage made from soybeans, caramel, sugar, and water, has the nutritional value of cow's milk. A tasteless, colorless fish protein concentrate, to be added to other foods, is being produced in Peru from fish not normally edible. Scientists in India have found an economical way to produce synthetic lysine, an amino acid. In the larger Indian cities, bakeries fortify their wheat flour with this product and produce what may be the most nutritious bread in the world. High-lysine corn has a great appeal to the large corn-eating regions of Latin America and the Sahel.

The Soviet Union is currently diverting and turning southward four rivers which empty into the Arctic Ocean. Their plan is to increase the available amount of irrigation water for greater food output. Sub-Saharan Africa and the Amazon Basin of Brazil are two regions where they are sizable portions of arable land which can be well-watered. Various nations have undertaken projects attempting to directly harness atmospheric moisture to either move or create rainfall. *Desalinization* projects like the one in Jeddah, Saudi Arabia, irrigate coastal deserts by converting sea water to a usable form. To date, this process has been very costly and largely unprofitable.

Most of the readily cultivated land in the world is already being occupied and farmed. In general, the best hope for the immediate future lies in increasing productivity. World average wheat yields in 1976 were slightly over 20 *cwt* per acre; some countries were able to average 30, some individual farms 60.

The chief organization charged with combating worldwide agricultural problems is the United Nations Food and Agricultural Organization (FAO). Founded in Quebec in 1945 to help those countries most depleted of resources by World War II, the FAO has been headquartered in Rome since 1951. The following is a condensation of the FAO constitution, outlining the commitment of the organization to promote and recommend national and international action with respect to:

1. Scientific, technical, social, and economic research relating to nutrition, food, and agriculture
2. The improvement of education and administration relating to nutrition, food, and agriculture, and the spread of public knowledge of nutritional and agricultural science and practice
3. The conservation of natural resources and the adoption of improved methods of agricultural production and land *reclamation*
4. The improvement of the marketing, processing, and distribution of food and agricultural products
5. The adoption of policies for the provision of adequate agricultural credit, both national and international
6. The adoption of international policies on agricultural commodity arrangements

It is also the function of the FAO:

7. To furnish such technical assistance as governments may request
8. To organize, in cooperation with the governments concerned, such missions as may be needed to assist them in fulfilling the obligations arising from their acceptance of the UN Conference on Food and Agriculture.

Considering its enormous task, the FAO is doing a commendable job. It sponsors and carries out the work of many programs such as the World Food Program, the World Food Congress, the Freedom from Hunger Campaign, and the Indicative World Plan for Agricultural Development. Since only about one person in four is really well-fed and adequately nourished, and another 25% are seriously and chronically undernourished, the tasks of these self-defined programs are staggering.

As the world food crisis becomes more acute, it is clear that the field of agriculture needs trained, skilled professionals. In addition to the farmers themselves, there is a need for technically trained people in many aspects of this vital industry.

Agricultural engineers apply basic engineering principles to the solution of specific agricultural problems. They work at finding more efficient ways to design farm machinery and buildings, to use water and fertilizer, to conserve water and soil, and to process products, increase yields, and improve quality.

Courtesy United Nations

An agronomist at the Argonne National Laboratory weighs birdsfoot trefoil plants to test water retention in a soil conservation project.

Agronomists work to improve crop quality and yields by studying climate, soil, and environmental factors affecting a particular crop or region.

Soil scientists are responsible for the analysis and classification of soils, the prediction of their behavior under various farming methods, and the recommendation of plants best suited to specific areas.

Soil conservationists provide technical information and advice based on knowledge of rainfall, drainage conditions, erosion potential, and planting schedules and patterns. Their primary duty is to aid farmers in developing crop rotation and planting methods, and in the conservation of both soil and water.

Plant pathologists work to control or eliminate diseases caused by bacteria, viruses, fungi, rust, dust, and other plant enemies.

Entomologists study insects in order to eliminate or control those harmful to plants and to make use of those beneficial to plants. Entomology also involves experimentation with the commercial uses of pollination and insect-produced substances such as wax.

Agricultural economists work mainly in the areas of production, financing, and marketing of food products by analyzing and interpreting economic factors.

Geneticists develop new hybrids, strains, and varieties to meet specific production, marketing, or social conditions.

Food technologists strive to improve the flavor, color, and texture of foods without reducing their nutritional value. Their work, particularly in the areas of baked goods and canned foods, often combines the sciences of chemistry, bacteriology, and biology.

Agricultural extension service workers distribute information and instruct farmers in the latest discoveries and techniques developed by the specialists. They demonstrate and advise on all phases of technologies and their applications, individually, in formal university classes, or at informal gatherings of farmers' organizations or cooperatives.

In 1975 a ruinous drought cut the corn crop of Honduras by 50%. The United States, Japan, Canada, and West Germany sent immediate food aid, but many people starved. There was an almost total lack of adequate storage facilities and an acute shortage of trucks. The result was that most of the food stayed in urban areas where it was least needed, rotted in the sun on the wharves, or was diverted for private profit. All of the technology, resources, and organization in our complex world was not enough to get the food to the people who were starving. Mahatma Gandhi once said: "To the millions who have to go without two meals a day, the only acceptable form in which God dare appear is food."

Discussion

1. What do the UN's population projections tell us? How will this affect the need for agriculture production?

2. What happens in freshwater bodies when great amounts of fertilizer are washed in from nearby fields?

3. Give two examples of environmental difficulties caused by agricultural activities.

4. What is an endangered species? Give some examples.

5. Why do some countries ban the use of DDT? Why do others allow its use?

6. What usually happens when marginal land is cleared for cultivation?

7. What happens when land loses its natural vegetation or grass cover?

8. What do you think of the statement: "Soil is only rock on its way to the sea?"

9. What is a dust bowl?

10. How can particulate matter affect our climate?

11. What is one of the ultimate consequences of a frost line moving toward the Equator?

12. Give some specific examples of the effects of the green revolution.

13. How is the protein value of soybeans being exploited?

14. What is lysine? How is it being used in India?

15. Give some examples of new attempts at irrigation.

16. What is the FAO? Why was it established?

17. What does the FAO do with regard to farm credit?

18. Do you know of any FAO publications?

19. What do agricultural engineers and agronomists have in common?

20. Is the world of a plant pathologist the same as that of an entomologist? Explain.

21. Where do agricultural extension service workers perform their duties?

Review

A. Discuss various career opportunities in the field of agriculture. Which appeals to you most? Why?

B. You are a wheat farmer with severe pest problems. Also, there seems to be too much leaching of nutrients from your soil, probably caused by poor drainage. To what specialists might you turn for advice?

C. Describe the activities of the FAO. Are there any FAO activities in your country? What type? Would you ever want to work for the FAO? Why?

D. Is there much undernourishment in your country? What is being done to combat it? By whom? How could you become involved in solving this problem?

E. What do you think the future holds for agriculture in your country?

INDEX OF SPECIAL TERMS

ACRE	13	FERTILIZER	14
AGRIBUSINESS	85	FLAIL	14
AGRICULTURE	1	FODDER	2
ALGAE-LADEN SWAMPS	95	FOOD ADDITIVES	85
AMINO ACIDS	71	FOOD PRESERVATIVES	85
ANNUALS	43	FORAGE	71
ARABLE	1	FOWLS	71
ARTIFICIAL INSEMINATION	72	FREEZING	85
ATMOSPHERE	95	FUNGICIDE	30
BIENNIAL	44	FURROW	13
BIOSPHERE	95	GENETICS	29
BLANCHING	86	GRADE	85
BREED	72	GRAIN	13
BULB	44	GRAZING	43
BUND	59	GREENHOUSE	44
CWT:	96	GREEN REVOLUTION	96
CARBOHYDRATES	43	HARROW	13
CASH CROPS	59	HARVEST	1
COMBINE	14	HAY	71
CONCENTRATES	71	HERBICIDE	30
CONFINEMENT	72	HOE	13
CROP	1	HYBRIDS	29
CULTIVATE	1	HYDROPONICS	59
CUTTING	1	INSECTICIDE	30
DDT	95	IRRADIATION	86
DEFORESTATION	95	IRRIGATION	1
DESALINATION	96	LAND REFORM	85
DOMESTICATE	1	LATERIZATION	95
DRAFT ANIMALS	13	LEACHING	29
DROUGHT	43	LEGUME	30
DRYLAND FARMING	59	LIVESTOCK	43
ELEMENTS	29	MACRONUTRIENTS	29
ENVIRONMENT	95	MANURE	30
EROSION	29	MARGINAL	95
EUTROPHICATION	95	MARKETING	85
FALLOW	2	MICRONUTRIENTS	30
FARMING	1	MILLING	13
FATS	43	MILPA	59
FERTILE	1	MINERALS	43

MOLASSES 71
MONOCULTURE 59
MONSOON 43
MOW 14
NOMADS 85
NUTRIENTS 29
NUTRITION
ORGANIC 30
PARASITE 72
PARTICULATES 95
PASTEURIZATION 86
PASTURE 71
PERENNIAL 43
PESTICIDE 30
PHOTOSYNTHESIS 44
PLANKTON 96
PLANTATION FARMING 59
PLOW 13
PROCESSING 85
PROTEINS 43
PUREBRED 72
QUARANTINE 72
RATION 71
REAP 2
REAPER 14
RECLAMATION 96
RESERVOIR 59
RHIZOME 44
ROUGHAGE 13
RUMEN 71
RUMINANT 71
SILAGE 71
SILT 2
SOIL 29
SOW 1
SUBSISTENCE FARMING 59
TERRACING 1
THRESH 2
TRACTOR 13
TRUSS 59
TUBER 44
VACCINE 72
VARIETY TRIALS 29
VITAMINS 43
WATER TABLE 59
WEEDING 13
WINNOW 14
YIELD 1

NOTES

NOTES

NOTES

NOTES

NOTES

NOTES